PRIMATE
RESEARCH

W0051147

FASEB MONOGRAPHS

General Editor: KARL F. HEUMANN

A Continuation Order Plan is available for this series. A continuation order will bring delivery of each new volume immediately upon publication. Volumes are billed only upon actual shipment. For further information please contact Plenum Press.

PRIMATE RESEARCH

Edited by
William J. Goodwin
and
James Augustine
National Institutes of Health
Bethesda, Maryland

Springer Science+Business Media, LLC

Library of Congress Cataloging in Publication Data

Main entry under title:

Primate research.

(FASEB monographs; v. 6)
"The material in this book originally appeared in Federation proceedings, vol. 34, No. 8, July, 1975."
Includes index.
1. Primates—Physiology. I. Goodwin, William J. II. Augustine, James. III. Series: Federation of American Societies for Experimental Biology. FASEB monographs; v. 6.
QL737.P9P673 1976   599'.8'041   76-14940

ISBN 978-1-4684-2642-7   ISBN 978-1-4684-2640-3 (eBook)
DOI 10.1007.978-1-4684-2640-3

The material in this book originally appeared in *Federation Proceedings* Vol. 34, No. 8, July, 1975. First published in the present form by Plenum Publishing Corporation in 1976.

Copyright ©1975. Springer Science+Business Media New York
Originally published by Plenum US in 1975

softcover reprint of the hardcover 1st edition 1975

All rights reserved

No part of this book may be reproduced, stored in a retrieval system, or transmitted, in any form or by any means, electronic, mechanical, photocopying, microfilming, recording, or otherwise, without written permission

Contents

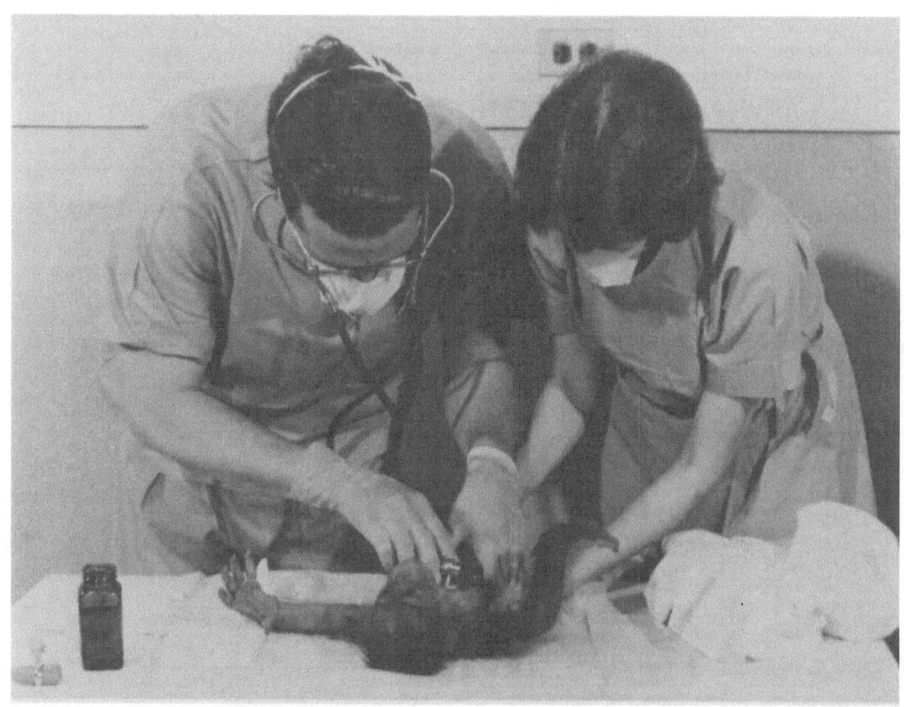

At the Yerkes Regional Primate Research Center, Atlanta, GA, Seriba, the center's firstborn orangutan, receives her initial physical exam. In the last 5 years, 21 baby orangutans have been born as part of the center's breeding program.

The Primate Research Centers Program of the National Institutes of Health

WILLIAM J. GOODWIN AND JAMES AUGUSTINE

Division of Research Resources, National Institutes of Health
Bethesda, Maryland 20014

The National Institutes of Health's Primate Research Centers Program was started in 1960. It was begun on the premise that research into human disease requires a close animal relative in which diseases could be duplicated and studied, their causes and effects documented, and effective methods of prevention and treatment developed. Within a decade, seven unique facilities had been established that were providing significant research data.

During early discussions and subsequent appropriation hearings, the U.S. Congress recognized the importance of establishing a national program of primate research facilities. Primates are expensive and difficult to maintain in individual research laboratories in sufficient numbers to provide meaningful data. Specialized facilities, well-trained professional and supporting personnel, and a variety of other distinctive resources are required.

When the program was established, it was decided that the centers were to receive not only initial funds but also continuing support from Federal sources. Planned as a long-range, coordinated effort in which many types of studies could be pursued in cooperation with major universities, the seven facilities are today supported by grants from NIH's Division of Research Resources.

Collectively, the seven centers are concerned with more than 900 different research projects in physiology, behavior, and disease. They maintain colonies of some 7,800 primates representing 48 species. Approximately 141 scientists work with 502 collaborating investigators and visiting scientists from all over the country and with 127 graduate students. The staffs of supporting personnel in the centers total 701. During one recent year, more than 730 scientific publications, including 53 books, were published as a result of the primate centers program.

These figures reflect the specific objectives of the program, which are:

* to pursue basic and applied biomedical research directed toward solution of human health and social problems.

* to establish a resource of scientists in many disciplines who are trained in the use of primates and can help maintain both continuity and high scientific quality in research.

* to provide opportunities for research and research training not otherwise available to visiting scientists, postdoctoral fellows, residents, junior faculty members, and to graduate, medical, dental, and veterinary students.

* to determine which problems of medical research are best pursued with nonhuman primates and which species are superior for particular studies.

* to develop improved breeding practices in order to increase the supplies of pedigreed, disease-free primates available for research.

* to study the natural diseases of primates and techniques of importation, conditioning, housing, and management which have an influence on the animal's well-being and suitability for studies.

* to develop new methods and equipment for primate studies.

* to supply biological specimens to qualified investigators.

* to disseminate the findings of studies done at the centers to primate users and others throughout the world.

Briefly stated, the major research emphasis and unique features of each of the seven centers supported by the program are as follows:

Oregon Regional Primate Research Center, Beaverton, Oregon. The major research activities of the Oregon Center's 45 scientists revolve around male and female reproductive biology, cardiovascular and metabolic diseases, immune diseases, and cutaneous biology. The center's unique features include a two-acre enclosure for behavioral studies on an intact troop of Japanese macaques, and the largest colony of prosimians found in the United States.

Washington Regional Primate Research Center, Seattle, Washington. This center has a significant program of research into central nervous system control of the cardiovascular system. In addition, the center has an extensive collaborative research program involving a larger number of scientists in many biomedical research disciplines. Also, a Primate Information Center supplies bibliographic information on biomedical research using primates. The center's breeding facility is located at Medical Lake, Washington, in a former maximum security mental institution.

Delta Regional Primate Research Center, Covington, Louisiana. The primary mission of the Delta Center is in infectious disease research which is carried out in the center's unique special isolation facility. The building offers such complete protection that only a few laboratories comparable to it can be found in the United States, and none at a primate center. Delta, which is located in a semi-tropical setting, is engaged in a number of primate breeding programs.

Yerkes Primate Research Center, Atlanta, Georgia. Originally located in Orange Park, Fla., the Yerkes Center was moved in 1965 to Emory University, to become an NIH primate research center. Housing the largest collection of great apes of any facility in the world, the mission of the center is in the neural and behavioral sciences and experimental pathology, including neoplastic and degenerative diseases. In addition to the main facility, a 115-acre field station is located 30 miles from Atlanta that has a number of outdoor facilities for great apes and monkeys.

Wisconsin Regional Primate Research Center, Madison, Wisconsin. The center has a long-standing emphasis on primate studies which relate closely to developmental problems in man. Along with behavorial research, reproductive physiology and neuro-physiology are other major research areas. A staff member is presently running a survey of primate populations in Africa.

New England Regional Primate Research Center, Southborough, Massachusetts. Located 30 miles from Boston, this center's primary research efforts are in microbiology and pathology. Though it has a relatively small core staff, it participates in a large collaborative program involving some 70 investigators from a variety of research areas. Emphasis is on the use of South American primates in biomedical research.

California Primate Research Center, Davis, California. This center was formerly known as the National Center for Primate Biology. Its research program includes studies on respiratory diseases, perinatal physiology, environmental toxicology and teratology, and infectious diseases. The center has a unique facility for studying the effects of air pollutants on the respiratory system.

Although each center is a distinct organizational entity, a host university of each maintains responsibility for its continuing administrative and scientific functioning. Financed by a budget of $11 million a year, grants are awarded by the Division of Research Resources to the host institution or to foundations which serve as grantee institutions.

As the use of primates in medical research continues to expand, the 11 papers which follow offer examples of the varied scientific activities in many areas of biomedical investigation currently under way at the seven centers.

Human melanoma and leukemia associated antigens defined by nonhuman primate antisera[1]

H. F. SEIGLER, R. S. METZGAR, T. MOHANAKUMAR, AND G. M. STUHLMILLER

*Yerkes Regional Primate Center, Emory University
Atlanta, Georgia 30322; Duke University Medical Center
Durham, North Carolina 27710; and the
Veterans Administration Hospital, Durham, North Carolina 27705*

Tumor cells express many of the antigens of their original host. The experiments conducted with transplantable tumors of rodents by early workers (2, 16) have demonstrated the multiplicity of antigens present on the cell surface. It has been elucidated that tumor cells share not only normal tissue antigens but that they may gain new antigenic specificities with malignant transformation (6, 9, 12, 14). These tumor associated antigens have been demonstrated to be both cross-reactive and individual, utilizing a variety of immunological techniques (3, 7, 13, 15). The presence of fetal or oncofetal antigens on tumor cell membranes has stimulated recent interest as well (1, 4).

Antibodies present in the sera of patients harboring a tumor, convalescent serum, and tumor eluates have not served as reliable sources of antibody reagents necessary for charac-terization of the antigenic mosaic of tumor cell membranes. Production of antisera to human tumor cells in rabbits and goats is complicated by a strong response of these species to human species-specific and allogeneic antigens. When antibodies to these normal cell antigens are finally absorbed out, the resultant antiserum is often weak or nonreactive with tumor-specific antigens. Since man and nonhuman primates share certain tissue alloantigens and have certain cross-reacting species anti-

[1] Supported by Public Health Service Grant RR 00165 from NIH to Yerkes Primate Research Center; NIH Grants AM0584 and CA 08975 to Duke University Medical Center and by NIH Grant AI 08897, Durham Hospital Program No. 7815-01 to the Veterans Administration Hospital.

Abbreviations: ALL, acute lymphocytic leukemia; AML, acute myelogenous leukemia; CLL, chronic lymphocytic leukemia, CGL, chronic granulocytic leukemia; HRBC, human erythrocytes; HWBC, human leukocytes.

gens, it was felt that simians would be better able to recognize the antigenic differences between normal and malignant human cells than would other mammalian species.

PRODUCTION OF CHIMPANZEE ANTISERA TO HUMAN MELANOMA CELL ANTIGENS

The chimpanzee has been successfully utilized to produce high titered specific cytotoxic HL-A alloantibodies (17). We have selected ABO compatible chimpanzees that share tissue alloantigens that cross-react with the HL-A phenotype of the immunizing tumor donor. The animal is immunized with tumor from a single donor so that a more specific antibody profile is realized. One may utilize fresh tumor cells, tumor cells maintained in tissue culture, or a purified solubilized tumor antigen extract. Both fresh tumor cells and cells maintained in tissue culture have been utilized. The sensitizing cell preparation and all subsequent boosters consisted of 2.5×10^7 to 2×10^8 melanoma cells suspended in sterile saline. One of the booster immunizations was done in an emulsion of Freund's complete adjuvant and the tumor cells. All subsequent boosters were done in each of the four extremities in the areas of erythema about the reactions to the adjuvant. The boosters were performed at 2-week intervals and each test bleeding was screened for antimelanoma activity after antiserum absorption with normal cells from the tumor donor. The absorptions were done at room temperature for 1 hour using 1×10^7 cells/ml. Antibody activity directed against normal antigens was completely removed with two consecutive absorptions. The chimpanzee antiserum was tested utilizing a modification of the microtechnique for complement de-

pendent cytotoxicity described by Mittal et al. (10). Preimmune chimpanzee serum is not cytotoxic for normal human lymphocytes, human fibroblasts, or for cultured melanoma cell lines.

ABSORPTION STUDIES UTILIZING AUTOLOGOUS AND ALLOGENEIC TUMOR CELLS

Absorption of the antiserum with normal lymphocytes and normal skin fibroblasts had little effect on the reactivity of the antiserum against the 14 melanoma target cells tested. Cytotoxicity activity to 1:16 or 1:32 dilution remained after absorption. Two sequential absorptions with 1×10^7 cells/ml with either fresh melanoma cells or melanoma cells maintained in tissue culture completely removed all activity of the antiserum when back tested against six different melanoma cell lines. In no instance could it be demonstrated that absorption with allogeneic melanoma cells left activity directed against autologous melanoma target cells. This chimpanzee antiserum, therefore, does not seem to detect a tumor associated antigen individual to the tumor host. It does, however, suggest that a cross-reacting tumor associated antigen for this tumor type does exist.

ABSORPTION STUDIES UTILIZING FETAL CELLS

Tissue culture cell lines derived from eight separate tissues of a 20-week-old fetus were utilized both as target cells and for absorption studies. The chimpanzee antiserum that had had all activity against normal tissues removed by absorption did, indeed, lyse fetal target cells. Absorption of the antiserum with normal cells and fetal cells removed all antifetal activity but left antibody directed to melanoma cells. Absorption of the antiserum with normal cells and melanoma cells removed all activity not only against the tumor target cells, but against the fetal cells as well.

FRACTIONATION AND ISOTOPE LABELING STUDIES

After absorption with normal cells the antiserum was fractionated using DEAE-Sephadex A-50 chromatography. The immunoelectrophoretically pure IgG fraction contained the cytotoxic activity directed against the melanoma cells. The IgG fraction was radiolabeled with ^{125}I utilizing both a chloramine T and a lactoperoxidase method. The antibody activity of the labeled antibody was somewhat reduced but maintained a good cytotoxic activity. If one labels the antibody at a concentration of one molecule of iodine per molecule of antibody, approximately 50% of the activity is lost. If this is reduced to 20% labeling, only one dilution is lost.

Lewis and Phillips have reported tumor associated antigens individual to the patient with melanoma (8). Others have demonstrated a cross-reactive antigen (3, 5, 13). The antiserum that we have raised in a chimpanzee demonstrates antibodies reactive with HL-A antigens, melanoma tumor associated antigen(s), and fetal or onco-fetal antigen(s). The antibody activity is concentrated in the IgG fraction and can be successfully radiolabeled. At present we are utilizing this antiserum to investigate serum blocking factors. This antibody also serves as the specific reagent for detecting the active antigenic extract in studies that we are conducting with solubilization and purification of the tumor antigens from cell membranes. These data will advance a more precise understanding of the antigenic profile present on tumor cell membranes. The antiserum is also being utilized in an in vitro lymphocyte dependent antibody test system in an effort to more clearly elucidate the host immune response to the malignant process. At the present time we are carrying out radioautographic techniques not only to demonstrate the specificity of the antibodies but to quantitate the binding affinity as well.

Other possible clinical uses of the antiserum include diagnostic radioscanning. If the labeled antibody is efficiently bound to the tumor it could serve as a specific isotope scan. The high titered cytotoxic antibody may also be valuable for passive serotherapy when administered alone or when complexed to chemotherapeutic agents and given as a combination immunochemotherapy regimen.

PRODUCTION OF SIMIAN ANTISERA TO HUMAN LEUKEMIA CELL ANTIGENS

Leukemia cells for immunization were obtained with an Aminco cell separator from peripheral blood of different leukemia patients. Test cells were prepared from peripheral blood leukocytes of normal adult volunteers, from patients with acute lymphocytic leukemia (ALL), acute myelogenous leukemia (AML), chronic lymphocytic leukemia (CLL), chronic granulocytic leukemia (CGL), and other neoplastic and nonneoplastic diseases. Lymphocytes or leukemic cell suspensions for cytotoxicity testing were made by removing adherent cell populations on nylon columns. Monkeys and chimpanzees were immunized with cells emulsified in Freund's complete adjuvant from individual HL-A phenotyped leukemic donors. Booster injections were administered without adjuvant into the edges of granulomas produced by the primary immunization. High titered (>1:80) leukemia specific, cytotoxic antibodies were detected after two booster injections. Subcellular materials, prepared by tryptic digestion of human cells from donors with acute myelogenous, or chronic lymphocytic or granulocytic leukemia, were also used as the immunizing antigens.

ABSORPTION AND CYTOTOXICITY TESTING OF SIMIAN ANTISERA TO HUMAN LEUKEMIC CELL ANTIGENS

For initial testing, the hyperimmune sera were heat inactivated and absorbed with human erythrocytes (HRBC) until the sera failed to agglutinate these cells. These antisera were then absorbed with human peripheral blood leukocytes (HWBC) from normal donors. Such absorbed antisera (HRBC and HWBC) were noncytotoxic to lymphocytes from all normal donors having various HL-A and W alloantigenic specificities. Aliquots of the antisera absorbed with human RBC and WBC were also absorbed with cells from individual leukemia donors in order to further establish their specificity. Cytotoxicity testing was performed by a modification (11) of the microtechnique described by Mittal et al. (10).

Monkey and chimpanzee antisera to human CLL cells after absorption with human RBC and WBC were cytotoxic to cells from all CLL and ALL patients tested. These antisera failed to react with cells from any of the normal donors or AML or CGL patients (Fig. 1—antisera M_1, M_2, C_1). Absorption of the anti-CLL sera with AML or CLG cells did not alter the reaction pattern of the antisera to CLL and ALL cells, but absorption with cells from individual CLL or ALL donors removed the antibody activity to all the CLL or ALL cells. These studies are in agreement with a concept that the leukemia associated antigens expressed on CLL or ALL cells are different from those on AML or CGL cells. Recent studies with nonhuman primate antisera to ALL cells, however, indicate that ALL cells also have an antigen that is not present on CLL cells.

The primate antisera to AML or CGL cells after absorptions with human RBC and WBC did not react with lymphocytes from normal donors or CLL or ALL patients, but were cytotoxic to cells from some but not all AML and CGL patients (Fig. 1—Antisera M_3, M_4, M_5, C_2). Cells from CLL and ALL patients failed to absorb the antibody activity of these sera to AML and CGL cells. The patterns of reactivity of the various myeloid antisera and data on absorption of the anti AML and CGL sera with cells from individual AML and CGL donors suggest a broad spectrum of antigens associated with myelogenous leukemia patients. Some of these antigens are cross-reactive between AML and CGL whereas a few others seem to be specifically associated with the individual type of leukemia (11). In addition, cells from some patients with myeloproliferative syndromes also react with the anti-AML and anti-CGL sera (11).

COMPARISON OF SPECIFICITY OF MONKEY AND RABBIT ANTISERA TO HUMAN LEUKEMIC CELL ANTIGENS

Subcellular material with leukemia specific antigen activity, as determined by inhibition of the cytotoxicity reactions of the nonhuman primate antisera, was prepared by trypsin digestion of cells from different leukemia donors. Monkeys and rabbits were then immunized with the same tryptic digest antigens from CLL,

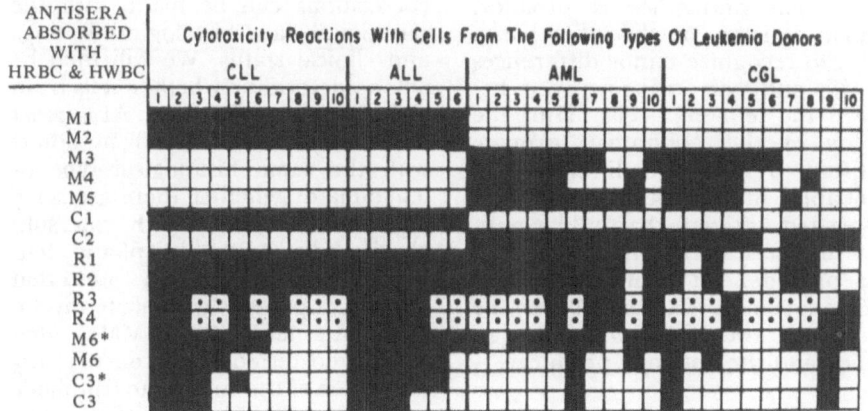

Figure 1. Cytotoxic reactions of various nonhuman primate and rabbit antisera to human and monkey leukemia cells. Antisera prepared in monkeys are designated M, chimpanzees C, and rabbits R. Antisera M1, M2, C1, R1 and R2 are to cells from donors with chronic lymphocytic leukemia. Antisera M3 and R4 to cells from donors with chronic granulocytic leukemia. Antisera M4, M5 and R3 are to cells from donors with acute myelogenous leukemia. Antisera M6 and C3 are to leukemic cells from monkeys.

CGL and AML donors. The monkey antisera to these antigens demonstrated the same serological specificity as that noted for simian antisera to whole leukemia cells. The rabbit antisera to tryptic digest antigens, however, demonstrated an altogether different serological pattern. All rabbit antisera, after absorptions with human RBC and WBC, failed to react with lymphocytes from any normal donors tested. However, cells from all types of leukemia patients, irrespective of their morphological classification, reacted with these antisera (Fig. 1—antisera R_1, R_2, R_3, R_4). Thus the rabbit antisera were unable to differentiate between the morphological classes of leukemias. Limited absorptions of these sera with cells from individual CLL, AML, and CGL donors indicated that the cytotoxic reactivity to cells from all leukemia donors can be absorbed by cells from any individual donor, whether CLL, AML, or CGL. There were also differences in the ability of the simian and rabbit antisera to react with cells from certain patients with other types of lymphoid malignancies. Antisera were, however, cytotoxic to lymphoblastoid cell lines established from human leukemia patients but failed to lyse an established cell line from a monkey myelogenous leukemia. In contrast, the rabbit antisera were cytotoxic to cells from some patients with lymphoma, Hodgkins, and lymphoproliferative disorders. These studies show that the rabbit and monkey antisera to antigens from the same leukemic cell donor are either reacting to different antigenic determinants of the same molecule or detecting different molecules of the leukemic cell membrane. In general, the serological specificity detected by the rabbit antibodies is broader. Simian antisera, on the other hand, seem to recognize minor differences in the antigenic makeup even between the leukemia cells. Thus the ability of the nonhuman primate antisera to recognize different and probably more specific leukemia associated antigens than rabbit antisera emphasizes the importance of the phylogenetic relationships and the antigenic differences between the species of the immunizing antigen donor and the antibody producer.

CLINICAL USE OF NONHUMAN PRIMATE ANTISERA TO HUMAN LEUKEMIA ASSOCIATED ANTIGENS

The primate antisera are already proving to be a valuable diagnostic tool. The antisera also can often distinguish the undifferentiated or unclassified leukemias as being either lymphatic or myeloid. In a few instances, patients with atypical leukemoid reaction of undetermined etiology have had reactivity with certain antisera to leukemia cells. The clinical significance of such findings is not clear but these patients are being closely followed serologically and clinically.

Our laboratories are now receiving blood samples from leukemia patients and patients with other hematologic disorders from many hospitals and medical centers in the eastern United States. The logistics of sample collection and shipment have been resolved and we are now able to do serological testing on heparinized, sterile, peripheral blood or bone marrow shipped by air mail special delivery, that arrive in our laboratory up to 48 to 72 hours after the sample was drawn. These collaborative studies are now providing the necessary numbers of patients of all ages with all types of leukemia so that better

correlations can be made with the serological data, histological findings and clinical status. We will then be able to determine whether cells from children with AML and ALL react differently from cells from adults with the same histological type of leukemia or whether there are serological differences which can subclassify the acute lymphatic leukemias. We have recently found that cells from patients with acute myelomonocytic leukemia (AMML) that have been tested react concurrently with some of the antisera to lymphatic leukemia and with some of the antisera to myeloid leukemia cells. Does this serological finding represent cross-reactivity or does this disease involve abnormalities of both cell types or their precursors?

The antisera have also been used for studies on the solubilization and isolation of the human leukemia associated antigens. Identification of the molecules involved will provide basic information for understanding the nature of the type of human malignant cell transformation. In addition, the purified antigens may provide better sources of antigen for immunotherapy studies than the current whole cell preparations which represent complex normal and malignant cell antigen mixtures.

SUMMARY

The rationale of our approach to the detection and characterization of human leukemia and melanoma associated membrane antigens is to take advantage of the unique immunological recognition perspective of simians to human antigens. There are enough antigenic similarities between man and nonhuman primates to blunt the immune response of apes and monkeys to human species-specific antigens. Their response to allogeneic normal human antigens such as HL-A can now be managed by in vitro absorption procedures. High titered antibodies specific for tumor associated membrane antigens have not been well documented in serological studies with patients' sera. Immunization of nonhuman primates utilizes a fully immunocompetent recipient who has no tumor and is not on chemotherapy. The immunization can utilize adjuvants and other regimens that morally and ethically cannot be easily used for human allogeneic immunization.

The simian antisera that are shown to be specific for human tumor membrane antigens can then be utilized for studies on the isolation and characterization of the antigens. Similar studies with HL-A antigens indicated that the solubilized partially purified antigens can then be used to elicit high titered antibodies in other simians. These latter antisera can then be used for more sensitive assays and studies on the association of the antigen(s) with the cell membrane.

Although the work summarized here covers leukemia and melanoma antigens, it is felt the approach will serve as a model for the detection by simian antisera of tumor-specific membrane antigens of cancer cells from patients with various types of solid tumors.

REFERENCES

1. BALDWIN, R. W., D. GLAVES AND B. M. VOSE. Differentiation between the embryonic and tumor specific antigens on chemically induced rat tumors. *Brit. J. Cancer* 29: 1, 1974.
2. BITTNER, J. J. A genetic study of transplantation of tumors arising in hybrid mice. *Am. J. Cancer* 15: 2202, 1931.
3. FOSSATI, G., M. I. COLNAGHI, G. DELLA PPORTA, N. CASCINELLI AND U. VERONESI. Cellular and humoral immunity against human malignant melanoma. *Intern. J. Cancer* 8: 344, 1971.
4. GOLD, P., AND S. O. FREEDMAN. Specific carcinoembryonic antigens of the human digestive system. *J. Exptl. Med.*
   122: 467, 1965.
5. GRAY, B. K., J. T. MEHIGAN AND D. L. MORTON. Demonstration of antibodies in melanoma patients' serum cytotoxic to human melanoma cells. *Proc. Am. Soc. Cancer Res.* 12: 79, 1971.
6. KLEIN, G. Tumor antigens. *Ann. Rev. Microbiol.* 20: 223, 1966.
7. LEWIS, M. G., R. L. IKONOPISOV, R. C. NAIRN, T. M. PHILLIPS, G. H. FAIRLY, D. C. BODENHAM AND P. ALEXANDER. Tumor-specific antibodies in human malignant melanoma and their relationship to the extent of the disease. *Brit. Med. J.* 3: 547, 1969.
8. LEWIS, M. G., AND T. M. PHILLIPS. Separation of two distinct tumor-associated antibodies in the serum of melanoma patients. *J. Nat. Cancer Inst.* 49: 915, 1972.
9. METZGAR, R. S., T. MOHANAKUMAR AND D. S. MILLER. Antigens specific for human lymphocytic and myeloid leukemia cells: Detection by nonhuman primate antiserums. *Science* 178: 986, 1972.
10. MITTAL, K. K., R. M. MICKEY, D. P. SINGAL AND P. I. TERASKI. Serotyping for homotransplantation. XVIII. Refinement of microdroplet lymphocytotoxicity test. *Transplantation* 6: 913, 1968.
11. MOHANAKUMAR, T., R. S. METZGAR AND D. S. MILLER. Human leukemia cell antigens: Serologic characterization with xenoantisera. *J. Natl. Cancer Inst.* 52: 1435, 1974.
12. MORTON, D. L. Immunological studies with human neoplasms. *J. Reticuloendothel. Soc.* 10: 137, 1971.
13. MORTON, D. L., R. A. MALMGREN, E. C. HOLMES AND A. S. KETCHAM. Demonstration of antibodies against human malignant melanoma by immunofluorescence. *Surgery* 64: 233, 1968.
14. OLD, L. J., AND E. A. BOYSE. Antigens of tumors and leukemias induced by viruses. *Federation Proc.* 24: 1009, 1965.
15. SEIGLER, H. F., W. W. SHINGLETON, R. S. METZGAR, E. C. BUCKLEY, P. M. BERGOC, D. S. MILLER, B. F. FETTER AND M. B. PHAUP. Non-specific and specific immunotherapy in patients with malignant melanoma. *Surgery* 72: 162, 1972.
16. STRONG, L. C. On the occurrence of mutations within transplantable neoplasms. *Genetics* 11: 294, 1926.
17. WARD, F. E., H. F. SEIGLER, R. S. METZGAR AND D. M. REID. Cross-reactivity of primate alloantigens: Absorption of anti-HL-A reactivity from human alloantisera by chimpanzee lymphocytes. *Tissue Antigens* 3: 389, 1973.

Cytogenetics of the squirrel monkey
(*Saimiri sciureus*)

T. C. JONES AND NANCY S. F. MA

New England Regional Primate Research Center
Harvard Medical School, Southborough, Massachusetts 01772

Squirrel monkeys occur in nature in widely separated geographic regions of Central and South America. These animals are of value in many aspects of biomedical research for which purposes they have been imported into the United States and Europe and are being reproduced in breeding colonies on both of these continents. Variations in the genetic and cytogenetic characteristics of these animals are of interest to the production and laboratory study of these animals as well as in their taxonomic classification. Our purpose in this paper is to review certain variations in the morphologic characteristics of the chromosomes of *Saimiri*.

PHENOTYPES AND GEOGRAPHIC ORIGIN

Several species and subspecies of squirrel monkeys have been tentatively assigned to the genus *Saimiri* but these classifications have been based on minor differences in coat coloration and are not clearly related to specific geographic regions (5). Napier and Napier (9) have assigned two species, *sciureus*, the "common squirrel monkey" and *oerstedii*, the

"red backed squirrel monkey," to the genus *Saimiri* and list eight subspecies. New information on the morphology of chromosomes and the geographic distribution of three specific types indicates that the taxonomic classification of these animals needs to be re-evaluated.

With careful observation, it is possible to distinguish the phenotypes that correspond to three presently known karyotypes, to be described, which appear to be found only in animals from certain geographic regions. It is difficult to determine the precise original habitat in the wild of squirrel monkeys that have been available for cytogenetic analysis, but we have associated three karyotypes with animals originating in specific areas of Central and South America. These regions may be identified by the name of the city from which animals are exported. A brief description of the observable features that may be used to distinguish these animals appears advisable.

Squirrel monkeys which have been collected in the vicinity of Iquitos, Peru may be recognized by the shape of the patches of white hairs which surround both eyes (Fig. 1). This

Figure 1. Female *Saimiri* exported from Iquitos, Peru. The white zone around the eyes bears sparse white hairs and the arch over the eyes is flattened ("roman arch").

feature may be described as two adjacent flattened ovoid areas which circumscribe each eye and coalesce at the midline. Two arches are thus produced by the white hair above each eye. In this phenotype the arches are distinctly flattened and have been compared to a "roman arch" by MacLean (8). The white hairs tend to be sparse around the eyes of these animals; the underlying skin thus gives this region a reddish tinge. Other characteristics of hair coat and body size are less consistent and of doubtful value in distinguishing this phenotype. The redbacked variety of Costa Rica and Panama may be distinguished by its coat color but its karyotype corresponds to the "Iquitos" type to be described.

The second phenotype that may be correlated to a distinctive karyotype is found in animals captured in the vicinity of Leticia, Colombia. The shape of the white zone around the eyes is the most useful detail to distinguish this type. The oval shape of

the two converging patches which surround each orbit is less flattened from the dorsal aspect. The arch formed above each eye is more conspicuous, sometimes reaching a point at the apex, and has been described as a "gothic arch" (8). This feature is illustrated in Fig. 2. The white hair tends to be more dense in these animals, therefore the unpigmented skin underlying them is less apparent. The white zone for this reason appears to be more intensely white.

The third distinguishable variety of squirrel monkey originates in Guyana and the animals are usually exported from Georgetown. The facial characteristics closely resemble the "Leticia" type, differing only in a few ways (Fig. 3). The white "eye patch" is arched more like the "gothic" type and the white hairs are usually quite dense. A roughly horizontal line, consisting of a few dark hairs, is usually evident at about the middle of the white zone above each orbit. The tips of the ears are often more

Figure 2. Male *Saimiri* captured in the vicinity of Leticia, Colombia. The white zone around the eyes contains dense white hair and the arch formed over the eyes reaches an apex ("gothic arch").

Figure 3. Female *Saimiri* from Georgetown, Guyana. The hair in the periocular zone is white and some black hairs are evident above the orbits. The ears bear many long coarse hairs.

inclined outward and the anterior surface of the pinna bears long coarse hairs.

It may be evident from the above descriptions that the differences between these phenotypes are subtle. We prefer to rely on chromosomal analysis to distinguish them.

THE KARYOTYPES

The modal diploid number ($2n = 44$,) of chromosomes is quite constant in all *Saimiri* that have been studied (5). The karyotype contains 21 pairs of autosomes plus the sex chromosome (XY) pair. Each chromosome may be identified by the G-banding technique (6). The chromosomes may be most conveniently studied by dividing them into three groups, based on centromeric position, then arranging the paired chromosomes in decreasing order, based on overall length, in each of these groups. Group A is made up of metacentric chromosomes in which the centromere is unquestionably located in

the middle of the chromosome. This group contains 5 pairs of chromosomes which may be designated by numbers A1 through 5 (Fig. 4). The number and morphology of this group of chromosomes has proved to be quite constant in our studies.

The second (B) group of chromosomes is made up of submetacentric or subtelocentric chromosomes in which the centromere is not located in the exact center of the chromosome. These have been designated B6 through 16 (Fig. 4). The total number of pairs in this group may be 11 (Iquitos), 10 (Leticia), or 9 (Georgetown), depending on the geographic origin and type, to be discussed.

The third (C) group of chromosomes in *Saimiri* are acrocentric, with the centromere near the end of the chromosome and little or no evident short arm. In the karyotypes studied so far, the number of acrocentric pairs were 5 (Iquitos), 6 (Leticia), or 7 (Georgetown).

The three karyotypes recognized so far among *Saimiri* are displayed in Fig. 4. The karyotypes in this figure are arranged in columns and each chromosome pair is grouped (A, B or C) and numbered in decreasing order of length. These chromosomes were prepared by the G-banding technique (6). *Column I* depicts the chromosomes of a female squirrel monkey imported from Iquitos, Peru and is representative of all animals in this population that have been studied. It will be seen that two acrocentric chromosomes C17 and C18 are not present in this karyotype but that submetacentrics B10 and B15 are represented. Specimens from animals originating in Costa Rica, Panama, and Pucallpa, Peru also conform to this karyotype.

The karyotype of a female *Saimiri* that was collected from the vicinity of Leticia, Colombia is displayed in column II of Fig. 4. Note that the B group of metacentric chromosomes (B1 through B5) is identical to those in column I. Chromosomes number

B15 (submetacentric) and C17 (acrocentric) are absent in this karyotype.

In the column I/II of Fig. 4, the karyotype is of a female squirrel

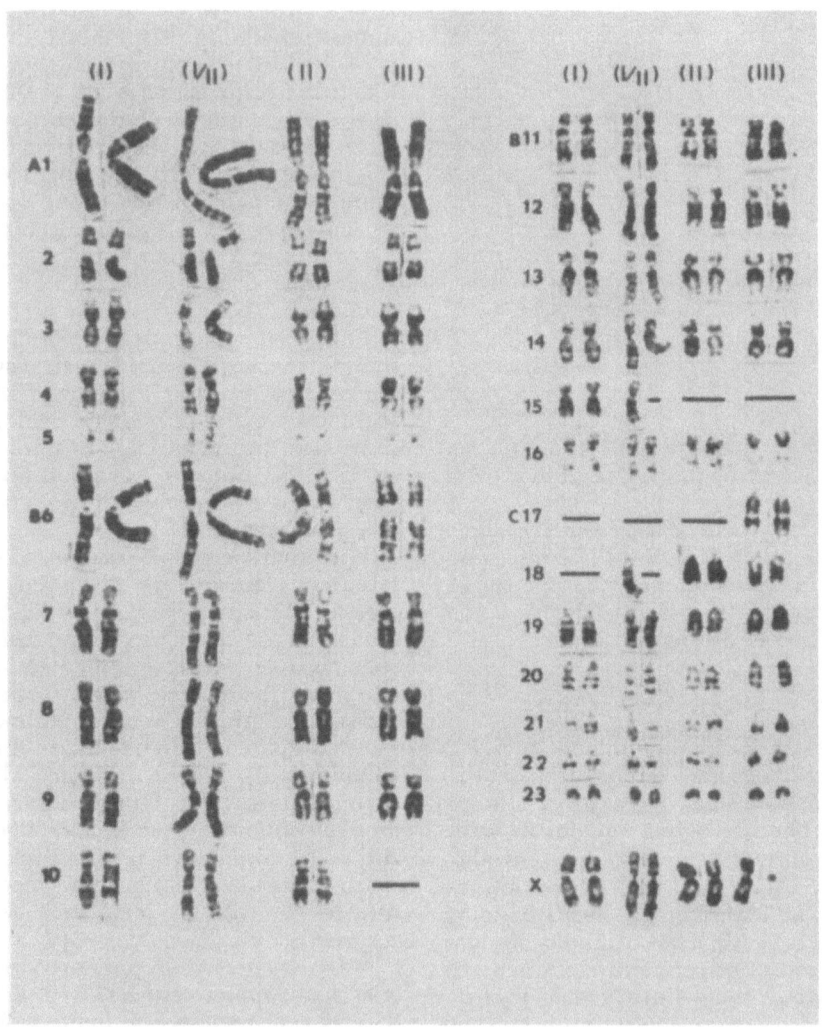

Figure 4. Chromosomes of *Saimiri* prepared by the G-banding technique. Karyotypes of four individual animals, representing different types and geographic origin, are arranged in vertical columns identified with Roman numerals. In *column I* is the karyotype of a female squirrel monkey of the "Iquitos" type. The karyotype of a "Leticia" type is presented in *column II*. In *column I/II* is the karyotype of a hybrid between the Iquitos and Leticia types. In *column III* is the karyotype of an animal from Georgetown, Guyana.

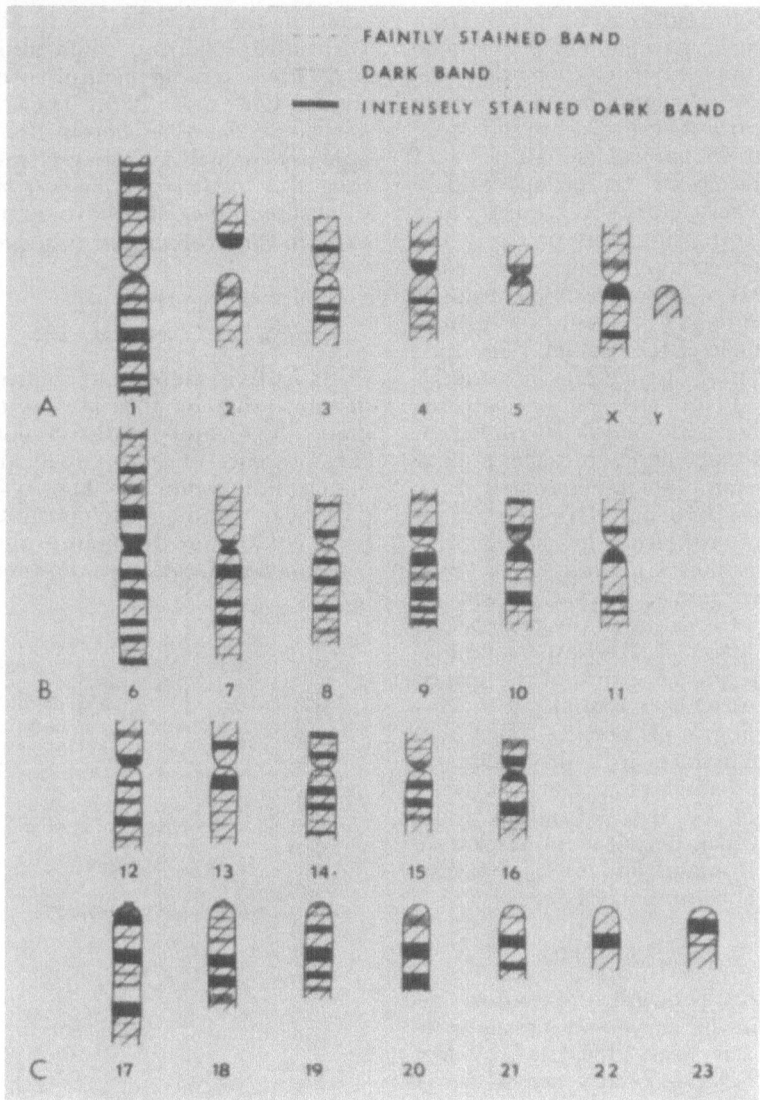

Figure 5. Idiogram of the chromosomes of *Saimiri* to illustrate the G-bands. This chart includes all of the presently known chromosomes of squirrel monkeys.

monkey resulting from the mating in our laboratory of a male *Saimiri* from Iquitos (karyotype I) with a female from Leticia (karyotype II). The karyotype of this hybrid animal is clearly the result of combining the haploid set of chromosomes from each parent. One submetacentric chromosome (B15) was acquired from the male parent and one acrocentric chromosome (C18) came from the female parent. Thus, the male parent had 11 pairs submetacentric (B group) chromosomes and 5 pairs acrocentric (C group) chromosomes; the female parent had 10 pairs of submetacentrics (B group) and 6 pairs of acrocentrics (C group). Their hybrid offspring had precisely the expected complement: 10.5 pairs (total 21) of B-group submetacentric and 5.5 pairs (11 total) of C-group, acrocentric chromosomes. The third karyotype, discovered in animals exported from Georgetown, Guyana, is displayed in *Column III* of Fig. 4. In this karyotype, a male animal in this example, submetacentric chromosomes B10 and B15 are absent but are replaced by acrocentric chromosomes C17 and C18. Thus, this Georgetown karyotype can be recognized readily by the presence of only 9 pairs of B group (submetacentric) and 7 pairs of C group (acrocentric) chromosomes.

An idiogram is presented in Fig. 5 which illustrates in a diagrammatic way the distinctive banding patterns, prepared by the G-banding technique, which permit identification of each chromatid and each presently known chromosome of *Saimiri*.

MECHANISMS

The most probable mechanism involved in the karyotype variations in *Saimiri* has been elucidated by Ma et al. (6). The occurrence of pericentric inversions during evolution of *Saimiri* seems to offer the best explanation of these observed variations. This possibility is illustrated in Fig. 6. A break in chromosome C17 with inversion of a broken segment and reattachment as depicted could result in a chromosome with the morphologic features of B10. This rearrangement could also occur in the opposite direction, B10 to C17, indi-

cated by the heavy arrows in Fig. 6. Pericentric inversion could also explain the rearrangement of chromosomes C18 and B15. The three karyotypes described herein in *Saimiri* appear to be stable within each group, thus this structural rearrangement of chromosomes may have occurred early in the evolution of this species.

CONSTITUTIVE HETEROCHROMATIN

Constitutive heterochromatin is demonstrable by the C-band technique (3, 6) and its distribution in chromosomes of *Saimiri* is diagrammatically illustrated in Fig. 7. This particular technique demonstrates a band of heterochromatin at the centromere of each chromosome al-

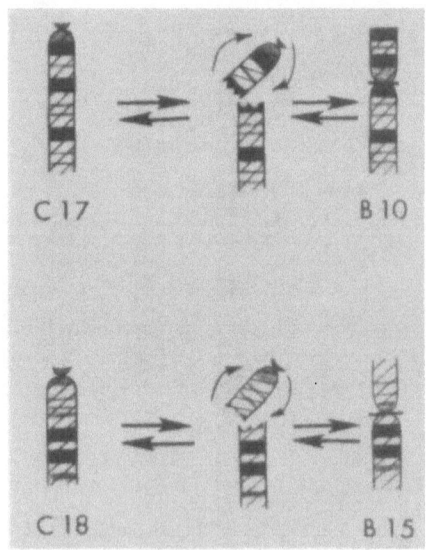

Figure 6. Mechanism most likely involved in the karyotype variation among *Saimiri*. Diagrammatic representation of pericentric inversion between chromosomes C17 and B10; C18 and B15. Small arrows symbolize the inversion that follows a break in the chromosome. Large arrows signify that the effect could occur in either direction.

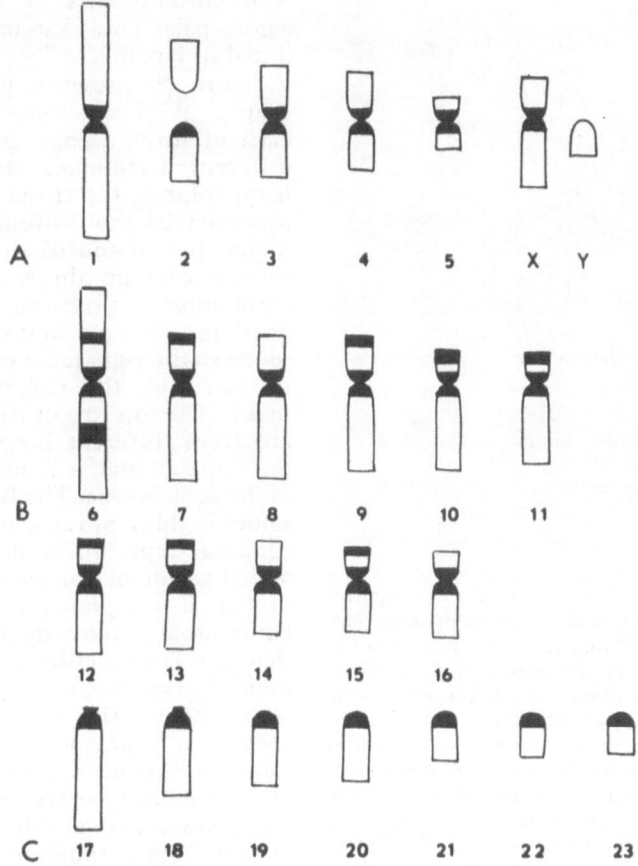

Figure 7. Diagrammatic representation of the constitutive heterochromatin of *Saimiri* as identified by the C-banding technique.

though sometimes this pericentromeric band may be absent in chromosome A2 (6). Several chromosomes (B7, B9, B10, B11, B12, B13, and B15) also have a band of heterochromatin at the telomeric end of the short arm. A band of heterochromatin is found within both arms (thus "interstitial") of one chromosome pair, number B6 (Fig. 7).

The heterochromatin bands found at the telomere of *Saimiri* chromosomes B10 and B16 have recently been shown to be useful in demonstrating an added segment in some animals (7). This variation, illustrated in Fig. 8, is of interest because the homozygous state, with added heterochromatin on both chromosomes of a pair, appears not to have been reported in other mammals although it is known to occur in chromosomes of some plants and insects. Chromosome B10 (Fig. 4) is an autosome that occurs in the karyotype of *Saimiri* of the Iquitos and Leticia phenotypes, but is replaced by the acrocentric chromosome C17 in the

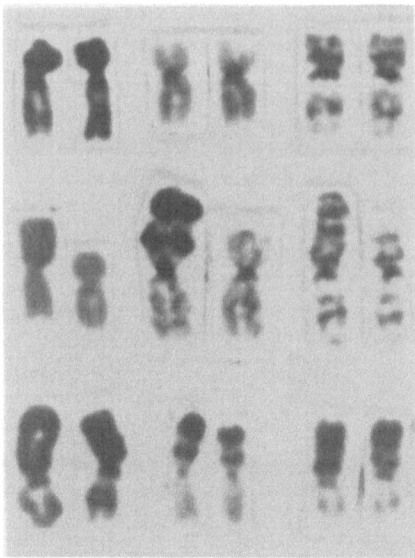

Figure 8. Chromosomes of three squirrel monkeys with added heterochromatin segments. The chromosomes are stained in the left column by the conventional technique, in the middle column by C-banding and in the right column by G-banding. The top row of chromosomes are number B16 from a squirrel monkey of the Iquitos type. No added segments are present in these. In the middle horizontal row are B16 chromosomes from a cross-bred monkey with one Iquitos and one Leticia parent, only one of each pair bears the added heterochromatin segment. Chromosomes B16 from an Iquitos type are illustrated in the bottom row. Both of these bear added heterochromatin on the short arm.

both chromosomes of this homologous pair. These variants are illustrated in Fig. 8.

Figure 8 presents photomicrographs of chromosome B16 from each of three *Saimiri*, prepared by different techniques. In the left hand column, the chromosomes are prepared by the conventional technique (not banded); the middle column contains three sets of B16 chromosomes prepared by the C-banding technique, and the chromosomes in the right hand column were prepared by the G-banding technique. The top row of chromosomes are from different preparations of the same animal, a squirrel monkey of the Iquitos type. The B16 chromosomes in this row represent the most common type which do not have added segments. In the middle row of Fig. 8 are photographs of B16 chromosomes from a squirrel monkey that is a cross between Iquitos and Leticia karyotypes. Only one chromosome bears the added heterochromatin segment. The chromosomes B16 from a squirrel monkey of the Iquitos type are illustrated on the bottom row of Fig. 8. In this animal, both chromosomes bear the added segment of heterochromatin on the short arm, which results in a change of this submetacentric pair to metacentrics. In conclusion, karyotypic differences in squirrel monkeys, *Saimiri*, stem mainly from the intrachromosomal rearrangement involving pericentric inversions and, secondarily, from accumulation of heterochromatin which increases the length of the short arm of certain chromosomes.

Georgetown type. Added heterochromatin on this chromosome was found in one chromosome of one monkey from Pucallpa, Peru and also in a second animal from Leticia. These were detected among 36 squirrel monkeys analyzed cytogenetically. In addition, among these same 36 animals, heteromorphic single chromosomes B16 were found in two. Chromosome B16 in one additional Iquitos-type animal had added heterochromatin on the short arm of

REFERENCES

1. BENDER, M. A., AND I. E. METTLER. Chromosome studies of primates. *Science* 128: 186–190, 1958.

2. DEBOER, L. E. M. Cytotaxonomy of the Platyrrhini (primates). *Genen Phaenen* 17: 1–115, 1974.
3. DRETS, M. E., AND M. W. SHAW. Specific banding patterns of human chromosomes. *Proc. Natl. Acad. Sci. U.S.* 68: 2073–2077, 1971.
4. EGOZCUE, J., E. M. PERKINS, F. HAGEMANAS AND D. M. FORD. The chromosomes of some Platyrrhini (*Callicebus, Ateles* and *Saimiri*). *Folia Primatol.* 11: 17–27, 1969.
5. JONES, T. C., R. W. THORINGTON, M. M. HU, E. ADAMS AND R. W. COOPER. Karyotypes of squirrel monkeys (*Saimiri sciureus*) from different geographic regions. *Am. J. Phys. Anthropol.* 38: 269–277, 1973.
6. MA, N. S. F., T. C. JONES, R. W. THORINGTON AND R. W. COOPER. Chromosome banding patterns in squirrel monkeys (*Saimiri sciureus*). *J. Med. Primatol.* 3: 120–137, 1974.

7. MA, N. S. F., AND T. C. JONES. Added heterochromatin segments of squirrel monkeys (*Saimiri sciureus*). *Folia Primatol.* In press.
8. MACLEAN, P. D. Mirror displays in the squirrel monkey, *Saimiri sciureus*. *Science* 146: 950–952, 1964.
9. NAPIER, J. R., AND P. H. NAPIER. *A Handbook of Living Primates.* New York: Academic, 1967, p. 309.
10. ROSENBLUM, L. A., AND R. W. COOPER. *The Squirrel Monkey.* New York: Academic, 1968, p. 1–29.
11. SRIVASTAVA, P. K., A. K. SRIVASTAVA AND F. V. LUCAS. Somatic chromosomes of squirrel monkey (*Saimiri sciureus*). *Primates* 10: 171–190, 1969.
12. WANG, H. C., AND S. FEDEROFF. Banding in human chromosomes treated with trypsin. *Nature New Biol.* 235: 52–53, 1972.

Fetal hormones and their effect on the differentiation of the central nervous system in primates[1]

JOHN A. RESKO

Reproductive Physiology, Oregon Regional Primate Research Center
Beaverton, Oregon 97005

Primate reproduction consists of two major and perhaps independent physiologic processes. First, the male and female are sexually attracted to each other for the purpose of coitus; second, pituitary gonadotropins are released which assure the deposit of fertile gametes in the reproductive tracts. In rodents, differences in the behavior of males and females and in the release of gonadotropins have been attributed to hormonal action during fetal (25) and neonatal development (2).

Male and female rodents differ in the predominant type of sexual posture they assume and the sexual partner they pursue. These behaviors can be changed if fetal genetic females are exposed at the appropriate time and dosage to the male hormone testosterone or if the testes of the male are removed at the appropriate time in development. If testosterone is given to rats soon after birth, the female is made anovulatory (2). Likewise if neonatal male rats are castrated, their sexual response as adults to estrogen and progesterone is like

that of females (4) and like females they respond to estrogen by releasing gonadotropin (24). Intact males, on the other hand, do not release gonadotropin after estrogen administration (24).

Primate sexual behavior is less hormone dependent than that of rodents (4). Although female primates copulate more frequently at midcycle and during the premenstrual period than at any other time, they continue to copulate throughout the intermenstrual period (21). In some species of primates, female sexual behaviors do not change with the ovarian cycle even though intromission rates and frequency of ejaculation change (9). Little is known about the action of fetal hormones on the development of the primate nervous system.

We have studied the function of steroid hormones in relation to sexual differentiation in primates. This work

---
[1] Publication No. 788 of the Oregon Regional Primate Research Center, supported by NIH Grants RR-00163, HD-07658, and HD-05969.

was divided into two phases: the first was descriptive, the second experimental. In the former, adult female rhesus monkeys (*Macaca mulatta*) were placed with breeding males on days 10, 11, and 12 of the menstrual cycle (day 1 being the first day of menstruation) for a breeding period of not more than 48 hr. If pregnancy (determined by palpation at 4 weeks) ensued, the gestational age of the fetus was computed from day 11 of the cycle. At various times between days 59 to 163 of gestation, pregnant females were taken to surgery and anesthetized (27), and the gravid uteri exposed by a midventral incision. The fetuses were then delivered by cesarean section with the umbilical cords intact. Depending on the age of the fetus, various quantities of blood were drawn from the umbilical vessels in heparinized syringes; for example, at 59 days, 0.8 to 1.0 ml of blood was drawn from the umbilical artery and then the vein; at 150 days and beyond, as much as 5 ml were drawn from each umbilical vessel. After centrifugation, the plasma was removed and stored at -16 C until steroid analyses were performed. Fetal gonadectomies were performed in some animals on day 100 of gestation (29). We measured several classes of steroid hormones—estrogens, androgens, and progestins—in plasma from the umbilical circulation of both intact and gonadectomized fetal rhesus monkeys by sensitive and specific radioimmunoassays (27, 29). The quantities of these hormones were compared in male and female fetuses. Our aim in this first phase was to describe the hormonal milieu in which the primate fetus develops.

The second phase was more experimental. Here we studied the effects of prenatal androgen on the hypothalamo-pituitary-gonadal axis of pregnant females bearing genetic female fetuses. The animals studied in this phase were pseudohermaphrodites produced by two of my colleagues, Drs. R. W. Goy and C. H. Phoenix. The treatment regimens, along with the times of treatment and the morphological alterations, have been reported (26).

ANDROGENS IN PLASMA OF FETAL RHESUS MONKEY

One consistent finding in these studies has been a sex difference in the amounts of testosterone found in the fetal circulation of the rhesus monkey (Fig. 1). Especially in the earlier states of gestation (<100 days), testosterone was higher in the fetal male than in the female. In this species, the fetal testis differentiates into a morphologically identifiable structure at about 39 days gestation. Our data do not include that stage, but at 59 days of age, male fetuses had more testosterone in their systemic circulation than females.

The source of the high levels of testosterone in the fetal male is most likely the fetal testis. These concentrations are significantly higher in the umbilical artery than in the umbilical vein (Fig. 2). No significant differences in the quantities of testosterone were found in the umbilical artery and vein of female fetuses (Fig. 2). In addition, testosterone levels in 3 males castrated on day 100 of gestation and sampled on days 150 to 156 were similar to those in females (Fig. 2). Fetal sex did not influence the level of this hormone in the maternal circulation.

A more careful analysis of the testosterone data indicates that the males divide arbitrarily into 3 categories (low, medium, and high) according to the quantities of testos-

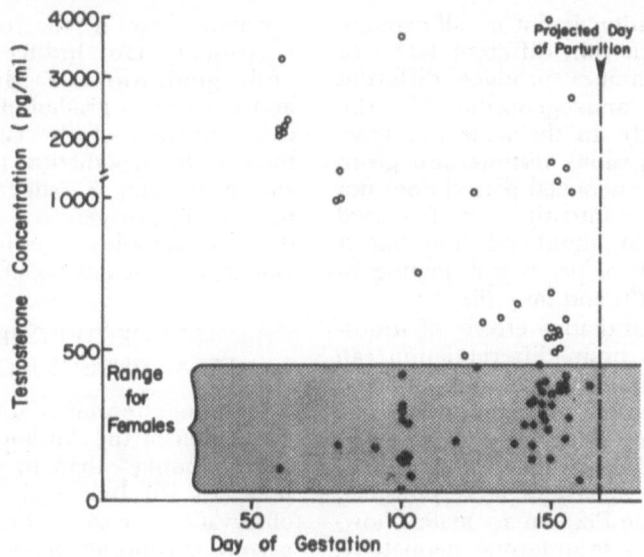

Figure 1. Testosterone concentrations in umbilical artery plasma of the fetal male (o) and female (•) rhesus monkey.

terone in their circulatory system (Fig. 3). At 100 to 155 days of gestation, some males had less than 500 pg/ml in their systemic circulation, i.e., values similar to those in female fetuses. Other fetuses (78 to 155 days of gestation) had medium, and still others (59 to 157 days of gestation) had high levels of testosterone. These hormone levels represent assays on blood samples drawn at single stages of gestation. We do not yet know how testosterone varies from one gestational age to another in the same animals especially in critical times for psychosexual development. This information, however, is important because if the levels are relatively constant, our data support the hypothesis that during fetal development some male fetuses are exposed to more testosterone than others.

Whether the amount of endogenous androgen to which the fetus is exposed is related to the degree of neural differentiation is not known.

Can quantitative differences in both sexual behavior and the release of gonadotropic hormones between members of the same sex be correlated with quantitative differences in steroid hormones present during fetal development? Establishing a causal relationship is complicated by the fact that: 1) There may be different thresholds of sensitivity to androgen in the developing anlagen of the central nervous system (CNS); and 2) the length of exposure to androgens is probably important for differentiation.

Examples from the literature on rodents may provide some basis for our speculations about the development of the CNS in primates. In the rat, 30 μg of testosterone given on postnatal day 5 produce an androgenizing effect within 12 hr (1). Smaller doses (0.5 μg to 0.05 μg) given twice daily produce the same effect but only after 10 days of treatment (30). There is evidence that differentiation is not an all-or-none process but that different levels of male hormones produce different levels of androgenization in the female (23). In the male rat, however, exogenous testosterone given during the neonatal period does not increase the intensity of male sexual behavior in adulthood (33) but it does produce precocious mating in male rats (3) and mice (6).

The quantitative effects of testosterone on brain differentiation can be best illustrated in the hamster. Males castrated in adulthood display female behaviors in response to exogenous estrogen and progesterone and male behaviors in response to testosterone (32). Intact males, however, given testosterone neonatally, respond behaviorally only to male

hormones, as if the testosterone treatment had induced greater androgenization. Evidence that androgen has a graded effect on the differentiation of the CNS of primates is still hypothetical, but the data shown in Fig. 3, suggesting that developing anlagen are exposed to different amounts of androgen, provide a basis for this possibility.

PROGESTERONE IN FETAL RHESUS MONKEY

More progesterone is found in the circulation of the developing female rhesus monkey than in that of the male (15, 19). In a replication of this observation, we showed that the progesterone concentrations in the fetal plasma of 33 male and 36 female

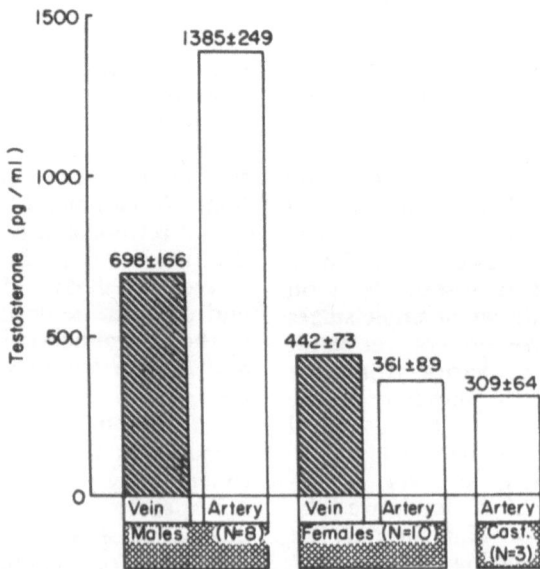

Figure 2. Testosterone in plasma from the umbilical vessels of the fetal rhesus monkey. Data from Resko et al. (29). *Statistics:* Male, testosterone in artery vs. vein: $P < 0.05$. Female, testosterone in artery vs. vein: $P > 0.05$. Artery, male vs. female: $P < 0.01$. Vein, male vs. female: $P > 0.05$. Artery, castrated males vs. intact: $P < 0.02$ (comparisons were made with seven intact males, 150 to 157 days of gestation). Three fetal males (labeled *cast.*) were castrated on day 100 and blood drawn from the umbilical artery on days 150 to 156 of gestation.

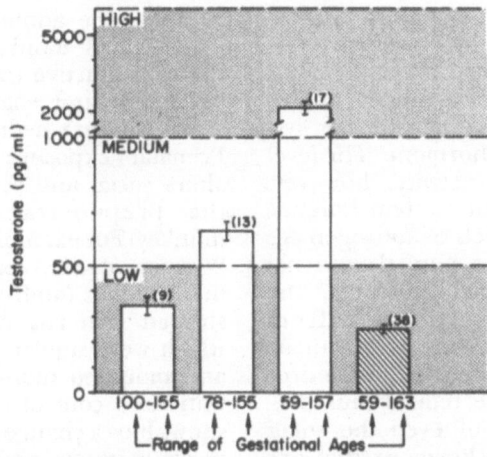

Figure 3. Arbitrary division (into low, medium, and high) of testosterone concentrations in the umbilical artery of fetal rhesus monkeys. Data are presented as means (bars) ± SE (vertical lines). Open bars are males; crosshatched is females. Numbers in parentheses are number of subjects that make up the respective means.

fetuses from 59 to 163 days of gestation were 4.10 ± 0.60 (SE) and 8.60 ± 1.10 (SE) ng/ml respectively; in other words, females had significantly more progesterone in their systemic circulation than males ($P < 0.01$) (Table 1). The significance of these observations is not known, but when viewed in the context of the antagonistic effects of progesterone on androgen action, they assume some importance. Under certain conditions, progesterone antagonizes the action of androgen, preventing the androgenizing action of testosterone in the neonatal rat (5, 18), androgen-induced aggressive behavior in the mouse (12), and sexual behavior in the guinea pig (7).

It has been postulated that progesterone exerts its antiandrogenic action by directly interfering with androgen action at the CNS (7, 10, 11), by inhibiting androgen production by the testes (20), or by inhibiting gonadotropin secretion (22). No one, to my knowledge, has shown that progesterone prevents androgen ac-

tion in the primate fetus, but neither has anyone shown that it cannot prevent this action. Therefore the hypothesis that hormonal differentiation of the reproductive system is not limited to the action of androgens alone but may involve other steroid hormones such as progesterone requires further study.

In Fig. 4 I have proposed a schema for sexual differentiation that involves the interaction of the two hormones testosterone and progesterone. According to this hypothesis, sexual differentiation is a consequence not of the action of androgen

TABLE 1. Progesterone in plasma[a] of the fetal rhesus monkey

Sex of fetus	No.	Progesterone, ng/ml ± SE
Male	33	4.10 ± 0.60[b]
Female	36	8.60 ± 1.10[b]

[a] Umbilical artery plasma from 59 to 163 days of gestation. [b] Concentrations of progesterone differed significantly between the sexes, $P < 0.01$.

alone but of the interaction of androgen with antiandrogen (progesterone). In the male fetus, the testis is active in the biosynthesis of testosterone and thus elevates the systemic pool of this hormone. The level of antiandrogen activity, however, is relatively low. In the female fetus, systemic blood levels of androgen are low because the ovary and the adrenal produce only small amounts, the remainder being derived from peripheral conversions of precursor substances. The fact that antiandrogen is high in the female thus prevents the action of even the small amounts of male hormone that are present. Changes in the quantities of either one of these hormones could alter sexual differentiation.

There is no doubt, however, that differentiation of the reproductive system of female primates can be altered by exogenous testosterone.

Testosterone administered to pregnant rhesus monkeys androgenizes the reproductive tract of the female fetus (33) and changes the central structures that mediate behavior (26). Prenatal exposure to testosterone alters social and sexual behavior in the prepubertal female rhesus monkey. For example, during the first 2 years of life, female pseudohermaphrodites (androgenized females) showed patterns of play behavior which were similar to those of males and mounted more frequently than untreated control females. Figure 5 shows how a change in function of the nervous system was effected by prenatal androgen treatment. In this figure, which compares the frequency of play initiation in males, females, and androgenized females, the mean frequency of play initiation in androgenized females is seen to differ significantly from that of their female

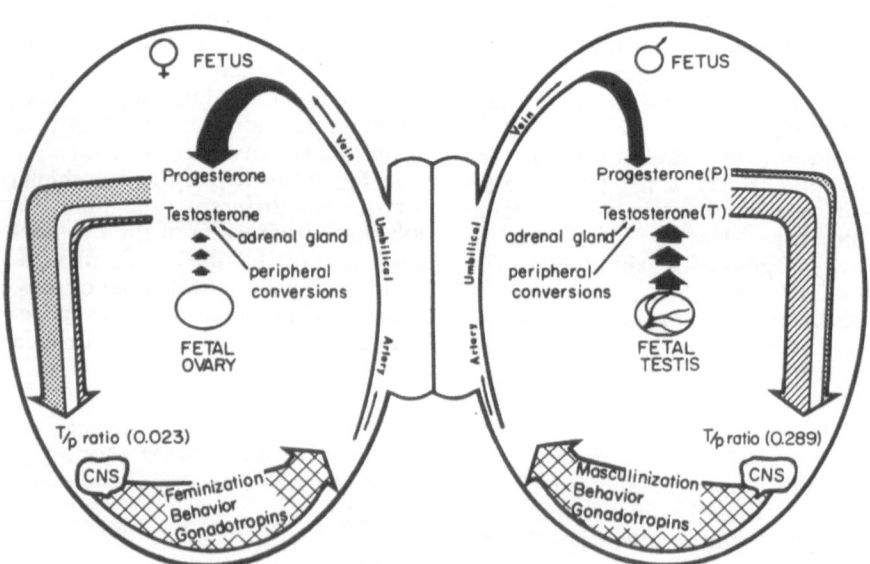

Figure 4. Possible interaction of steroid hormones on the differentiation of the central nervous system in the fetal rhesus monkey. The widths of the shaded sections of the arrows represent the quantities of progesterone (stippled areas) and testosterone (crosshatched areas) found in the fetal circulation. Taken from Resko (28).

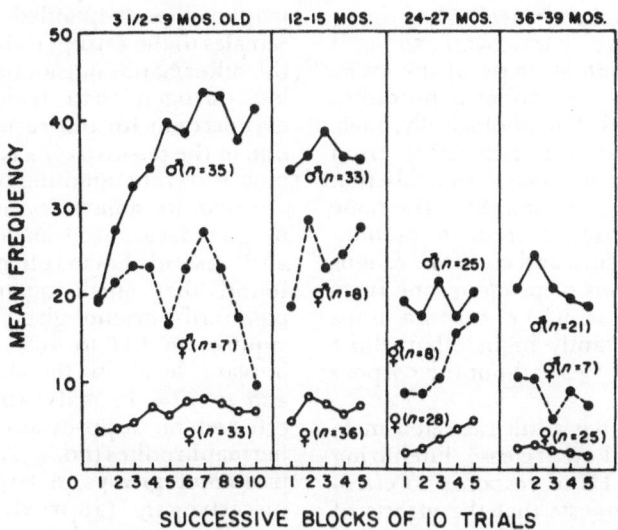

Figure 5. The frequency of performance of play initiation shown by males (● — ●), females (o — o), and pseudohermaphroditic female (● – – – ●) rhesus monkeys during the first 39 months of life. Taken from Goy and Resko (13).

counterparts (controls) and to approach that of infant males.

Other differences were observed in untreated and pseudohermaphroditic rhesus monkeys at the time of puberty (Fig. 6). In androgenized females the average age at menarche was 36.8 months (range 30.4–40.1, no. = 9); in untreated females, 29.2 months (range 21.3–34.7, no. = 18). These data suggest that in primates the mechanism that controls the onset of puberty is preset during fetal development and is androgen dependent. This observation should not be considered unusual since boys usually reach puberty later than girls; the association of this delay in the male with male hormones, however, was not clear.

In adulthood, androgenized female rhesus monkeys show more aggressive behavior than female controls (8). The fact that these behavioral differences continue into adulthood attests to the permanency

of androgen action on the neural structures during fetal development.

ACTION OF PRENATAL ANDROGENS ON THE NEURAL CENTERS WHICH REGULATE GONADOTROPIN

Like normal females, androgenized female monkeys go through a period of adolescent sterility after they reach menarche. After an initial period of endocrine adjustment, they begin to cycle regularly and to show signs of ovulation and the formation of a corpus luteum. Figure 7 presents the data for systemic concentrations of estradiol and progesterone throughout the intermenstrual period in representative groups of animals, one control and one experimental. The genetic females with androgenized external genitalia and modified behavior had a preovulatory surge of estradiol and levels of progesterone which indicated that ovulation had taken place. When these animals were later ovariectomized during the luteal phase of the cycle, 5 of the 8 pseudohermaphrodites had ovulated. Morphologically, their corpora lutea did not differ from those of the untreated controls that had been ovariectomized at the same time. In addition, in both pseudohermaphrodites and controls, venous concentrations of progesterone from the ovaries with the corpora lutea were significantly higher than those from the ovaries without the corpora lutea.

The fact that adult castrated male rhesus monkeys release luteinizing hormone (LH) in response to estrogen (16) suggests that the effects of androgen on the differentiation of the neural structures that mediate gonadotropin release are different in primates and in rodents and that the gonadotropin centers are independent of those areas of the brain that mediate behavior. Apparently the machinery for the so-called "positive feedback" effects of estrogen is present in the CNS of the male. In our laboratory, Dr. Robert Steiner repeated the work of Karsch et al. (16) and extended their observations to include the effects of exogenous estrogen on LH release in pseudohermaphroditic females (31). In castrate males, females, and androgenized females, LH levels were suppressed with a 1.5 cm silastic capsule which contained estradiol. An implant this size elevates the systemic concentration of estradiol to 53.0 ± 7.3 (SE) pg/ml (no. = 21). Within 48 hr after the single subcutaneous injection of estradiol benzoate in oil (50 µg/kg), significant elevations of LH were observed in all three groups. Our experiment and that of Karsch and co-workers differed in one respect, however. In ours, males and pseudoher-

maphrodites responded better than females to the estrogen challenge. On the other hand, our animals received less estrogen than theirs and this may account for our results. In addition to the positive effects, we studied how different amounts of estrogen affected its ability to suppress LH in gonadectomized males, females, and pseudohermaphrodites. We found that small quantities (53.0 pg/ml) of estradiol given for 30 days suppressed LH to 40% of the preimplant level in females (no. = 7) and to 77% · in males (no. = 7) and effected no suppression in pseudohermaphrodites (no. = 7). In all three treatment groups, a larger amount of estrogen (approximately 100 pg/ml) suppressed LH significantly from pretreatment levels. These data suggest that the hypothalamohypophyseal system which controls the secretion of LH in males and females differs in sensitivity to both the "positive" and negative feedback action of estradiol and that this sensitivity is partly a function of the prenatal hormonal environment. These differences were found in primates

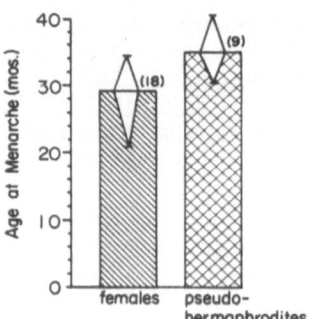

Figure 6. Age at menarche of untreated (hatched bar) and androgenized (crosshatched bar) female rhesus monkeys. Data are presented as means (bars) and ranges (spindles). Numbers in parentheses are numbers of subjects. Data redrawn from Goy and Resko (13).

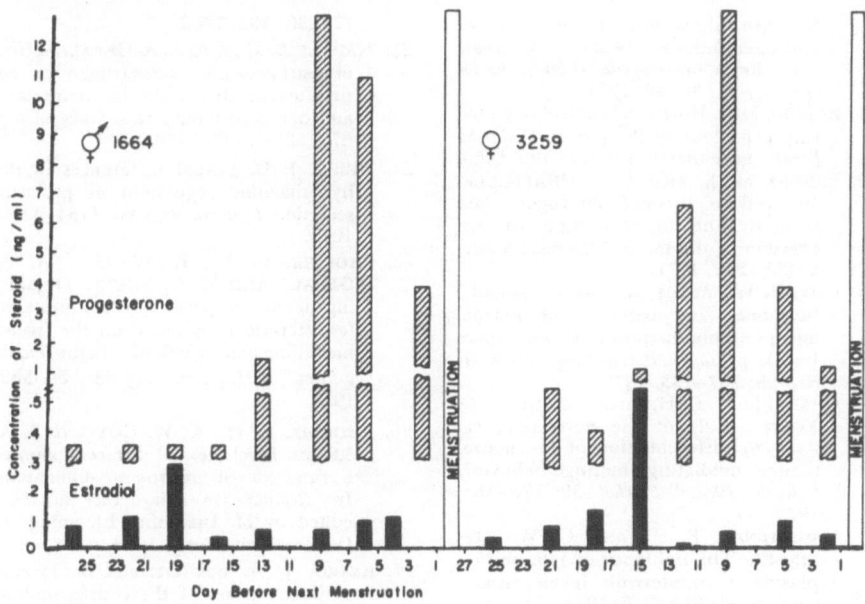

Figure 7. The concentrations of estradiol (solid bar) and progesterone (hatched bar) in systemic plasma of an untreated and a pseudohermaphroditic female rhesus monkey (cycle length = 27 days). Taken from Goy and Resko (13).

but not in rodents (24) and sheep (17). In these two species, exogenous estrogen does not facilitate the release of LH in males. What this species difference means to primates can only be determined by future experimentation.

REFERENCES

1. ARAI, Y., AND R. A. GORSKI. Critical exposure time for androgenization of the developing hypothalamus in the female rat. *Endocrinology* 82: 1010–1014, 1968.
2. BARRACLOUGH, C. A. Production of anovulatory, sterile rats by single injections of testosterone propionate. *Endocrinology* 68: 62–67, 1961.
3. BAUM, M. J. Precocious mating in male rats following treatment with androgen or estrogen. *J. Comp. Physiol. Psychol.* 78: 356–367, 1972.
4. BEACH, F. A. Human sexuality and evolution. In: *Reproductive Behavior Advances in Behavioral Biology*, edited by

W. Montagna and W. A. Sadler, New York: Plenum, 1974, vol. 11, p. 333.
5. CAGNONI, M., F. FANTINI, G. MORACE AND A. GHETTI. Failure of testosterone propionate to induce the "early androgen" syndrome in rats previously injected with progesterone. *J. Endocrinol.* 33: 527–528, 1965.
6. CAMPBELL, A. B., AND T. E. MCGILL. Neonatal hormone treatment and sexual behavior in male mice. *Hormone Behav.* 1: 145–150, 1970.
7. DIAMOND, M. Progesterone inhibition of normal sexual behavior in the male guinea pig. *Nature* 209: 1322–1324, 1966.
8. EATON, G. G., R. W. GOY AND C. H. PHOENIX. Effects of testosterone treatment in adulthood on sexual behaviour of female pseudohermaphrodite rhesus monkeys. *Nature (New Biol.)* 242: 119–120, 1973.
9. EATON, G. G., AND J. A. RESKO. Ovarian hormones and sexual behavior in *Macaca nemestrina. J. Comp. Physiol. Psychol.* 86: 919–925, 1974.
10. ERICKSON, C. J., R. H. BRUDER, B. R. KOMISARUK AND D. S. LEHRMAN.

Selective inhibition by progesterone of androgen-induced behavior in male ring doves (*Streptopelia risoria*). *Endocrinology* 81: 39–44, 1967.

11. ERPINO, M. J. Hormonal control of courtship behaviour in the pigeon (*Columba livia*). *Animal Behav.* 17: 401–405, 1969.

12. ERPINO, M. J., AND T. C. CHAPPELLE. Interactions between androgens and progesterone in mediation of aggression in the mouse. *Hormone Behav.* 2: 265–272, 1971.

13. GOY, R. W., AND J. A. RESKO. Gonadal hormones and behavior of normal and pseudohermaphroditic nonhuman female primates. *Recent Progr. Hormone Res.* 28: 707–733, 1972.

14. GRADY, K. L., C. H. PHOENIX AND W. C. YOUNG. Role of the developing rat testes in differentiation of the neural tissues mediating mating behavior. *J. Comp. Physiol. Psychol.* 59: 176–182, 1965.

15. HAGEMENAS, F. C., AND G. W. KITTINGER. The influence of fetal sex on plasma progesterone levels. *Endocrinology* 91: 253–256, 1972.

16. KARSCH, F. J., D. J. DIERSCHKE AND E. KNOBIL. Sexual differentiation of pituitary function: Apparent difference between primates and rodents. *Science* 179: 484–486, 1973.

17. KARSCH, F. J., AND D. L. FOSTER. Sex difference in the mechanism controlling the surge of luteinizing hormone in sheep. 56th Ann. Meeting Endocrine Society, Atlanta, Georgia, A-196, Abstract 281, 1974.

18. KINCL, F. A., AND M. MAGUEO. Prevention by progesterone of steroid-induced sterility in neonatal male and female rats. *Endocrinology* 77: 859–862, 1965.

19. MACDONALD, G. J., K. YOSHINAGA AND R. O. GREEP. Serum progesterone values in *Macaca mulatta* near term. *Am. J. Phys. Anthropol.* 38: 201–206, 1973.

20. MATSUMOTO, K., D. K. MAHAJAN AND L. T. SAMUELS. The influence of progesterone on the conversion of 17-hydroxyprogesterone to testosterone in the mouse testis. *Endocrinology* 94: 808–814, 1974.

21. MONEY, J., AND A. A. EHRHARDT. *Man and Woman Boy and Girl.* Baltimore: Johns Hopkins Univ. Press, 1972, chapt. 11, p. 222.

22. MURTON, R. K., R. J. P. THEARLE AND B. LOFTS. The endocrine basis of breeding behavior in the feral pigeon (*Columba livia*). I. Effects of exogenous hormones on the pre-incubation behavior of intact males. *Animal Behav.* 17: 286–306, 1969.

23. NAPOLI, A. M., AND A. A. GERALL. Effect of estrogen and antiestrogen on reproductive function in neonatally androgenized female rats. *Endocrinology* 87: 1330–1337, 1970.

24. NEILL, J. D. Sexual differences in the hypothalamic regulation of prolactin secretion. *Endocrinology* 90: 1154–1159, 1972.

25. PHOENIX, C. H., R. W. GOY, A. A. GERALL AND W. C. YOUNG. Organizing action of prenatally administered testosterone propionate on the tissues mediating mating behavior in the female guinea pig. *Endocrinology* 65: 369–382, 1959.

26. PHOENIX, C. H., R. W. GOY AND J. A. RESKO. Psychosexual differentiation as a function of androgen stimulation. In: *Reproduction and Sexual Behavior*, edited by M. Diamond. Bloomington: Indiana Univ. Press, 1968, p. 33–49.

27. RESKO, J. A. Sex steroids in adrenal effluent plasma of the ovariectomized rhesus monkey. *J. Clin. Endocrinol. Metab.* 33: 940–948, 1971.

28. RESKO, J. A. The relationship between fetal hormones and the differentiation of the central nervous system in primates. In: *Reproductive Behavior Advances in Behavioral Biology*, edited by W. Montagna and W. A. Sadler. New York: Plenum, 1974, vol. 11, p. 211.

29. RESKO, J. A., A. MALLEY, D. BEGLEY AND D. L. HESS. Radioimmunoassay of testosterone during fetal development of the rhesus monkey. *Endocrinology* 93: 156–161, 1973.

30. SHERIDAN, P. J., M. X. ZARROW AND V. H. DENENBERG. Androgenization of the neonatal female rat with very low doses of androgen. *J. Endocrinol.* 57: 33–45, 1973.

31. STEINER, R. A., D. K. CLIFTON, H. G. SPIES AND J. A. RESKO. Feedback control of LH by estradiol in female, male, and female pseudohermaphroditic rhesus monkeys. 56th Ann. Meeting Endocrine Society, Atlanta, Georgia, A-195, Abstract 280, 1974.

32. SWANSON, H. H. Determinations of the sex role in hamsters by the action of sex hormones in infancy. In: *The Influence of Hormones on the Nervous System*, edited by D. H. Ford. Basel: Karger, 1971, p. 424–440.

33. WELLS, L. J., AND G. VAN WAGENEN. Androgen-induced female pseudo-

hermaphroditism in the monkey (*Macaca mulatta*): Anatomy of the reproductive organs. In: *Carnegie Inst. Wash. Publ. 235; Contrib. Embryol.* 35: 95–106, 1954.

34. WHALEN, R. E. Differentiation of the neural mechanisms which control gonadotropin secretion and sexual behavior. In: *Perspectives in Reproduction and Sexual Behavior*, edited by M. Diamond. Bloomington: Indiana Univ. Press, 1968, p. 303–340.

Immunology of borreliosis in nonhuman primates[1]

OSCAR FELSENFELD AND ROBERT H. WOLF

Tulane University Delta Regional Primate Research Center
Covington, Louisiana 70433

Nonhuman primates have been found susceptible to infection with borrelias which cause relapsing fever in man (2). Preliminary experiments showed that patas monkeys (*Erythrocebus patas*) and vervets (African green monkeys, *Cercopithecus aethiops*) may serve as satisfactory models for the investigation of the immunologic features of infections with *Borrelia turicatae*[2] and *B. hermsii*, which are the two most frequently reported causative agents of human borreliosis in the United States. The antigenic analysis of the strains (6), the relationship of ribosomal activity to lymphoid tissue cytology (4), and the influence of the antigenic variations of borrelias on antibody formation in nonhuman primates have already been published (5).

The purpose of this communication is to correlate previous observations with hitherto unpublished data.

MATERIALS AND METHODS

Borrelia strains

B. turicatae Kansas strain was received in *Ornithodoros turicata* ticks. No antigenic variation of the strain was observed in the ticks and the ticks fed readily on monkeys. When infection with a precise number of borrelias was desired, the *Ornithodoros* were allowed to feed on infant mice from which heparinized blood was collected 3 ± 1 days after infection at the peak of the borrelemia. The borrelias were collected by centrifuging the blood specimens at $100 \times g$ for 5 min, then centrifuging the supernate at 4,000 to $5,000 \times g$ for 1 hr at 4 C. The borrelias in the second sediment were suspended in 0.01 M phosphate buffer pH 7.4 with 0.14M NaCl and 5 mM each of $CaCl_2$ and $MgCl_2$ (PBSCM) and counted in the Levy-Neubauer chamber.

B. hermsii was isolated from the blood of a patient who visited the Californian High Sierra. The strain displayed a predominantly biphasic variation and could be maintained by

[1] Partly supported by grant AI 10065 of the National Institutes of Health, Bethesda, MD.

[2] We are greatly indebted to Dr. Willy Burgdorfer, Rocky Mountain Laboratories, Public Health Service, for the *Borrelia turicatae* strain.

Abbreviations: PBSCM, 0.01M phosphate buffer pH 7.4 with 0.14 M NaCl, 5 mM $CaCl_2$ and 5 mM $MgCl_2$; IAC, immunologically active cells; IHA, indirect hemagglutination test.

successive mouse transfers without a statistically significant (<5%) appearance of intermediate phases. Recovery from infected mice was by the same method as *B. turicatae*.

Nonhuman primates

Adolescent and young adult primates were used. The patas weighed 3.4 to 6.7 (average 4.2) kg; the vervets 2.5 to 5.3 (average 3.9) kg. The animals were of either sex, and in apparently good physical condition on physical, hematologic, stool and bacteriologic examination. The animals were kept in strict isolation.

Serum specimens were obtained 7 days before, on the day of infection, and on days 3, 7, 14, 30 after infection. *Borrelia* counts were made in the peripheral blood daily prior to and during borrelemia. The examination continued at 2 to 3 day intervals for 2 wk after borrelemia if the animal remained alive. If a relapse occurred, daily *Borrelia* determinations were resumed. The smears of peripheral blood were examined by dark field microscopy.

Lymph nodes were excised under aseptic conditions and general anesthesia. The nodes were biopsied 7 days before and on days 7, 14, and 30 after infection. The lymph nodes were divided into three parts, one of which was frozen immediately by the isopentane-liquid N_2 method for fluorescent antibody and other cytologic studies. The second portion was ground in the Bronwill mill at 8,000 cycles with CO_2 cooling and frozen at -20 C for biochemical and serologic examinations. The third part was fixed in Carnoy's fluid for histologic studies.

The outcome of the testing of 2 patas which had more than one attack of borrelemia and of 8 which died after *B. turicatae* infection have been published (4, 5). They are not

included in this report. Therefore this communication contains only the results of the examination of 10 patas infected with *B. turicatae*. Ten of the 20 patas infected with *B. hermsii* succumbed. The results of only the immunologic studies of the surviving 10 patas are described here. Five of 10 vervets infected with *B. turicatae* survived. Sixteen vervets had to be infected and studied to accumulate data from 5 animals living for at least 30 days. The other 11, when sub finem, were treated with tetracycline and recovered fully. Their immunologic responses were described elsewhere (paper in print).

Isolation of immunoglobulins and preparation of antiserums

Immunoglobulins of the IgG and IgM classes were prepared from pools of serums of healthy patas and vervets according to the method of Reif (10). IgA was isolated by the method of Fine and Steinbach (7). All column effluents containing an Ig class were dialyzed against powdered sugar, lyophilized, or kept at -20 C without a preservative.

The immunoglobulins were tested for purity by immunoelectrophoresis and in the Ouchterlony test against rabbit whole anti-patas and anti-vervet serums.

The Ig solutions were adjusted to contain 5 mg/ml PBSCM. Rabbits were immunized according to the method of Uhr et al. (13). The serums were tested for purity and specificity by immunodiffusion and immunoelectrophoresis.

Immunologically active cells

Immunologically active cells (IAC) forming immunoglobulins were enumerated in the lymph node section by the indirect fluorescent antibody method of Crabbé and Heremans

(1). The IAC (prolymphocytes and proplasmacytes) were counted using a Leitz fluorescent microscope. Frozen sections without one of the reagents and parallel sections stained according to Unna-Pappenheim served as controls.

Determination of immunoglobulins in serums and lymph nodes

A modification of the radial immunodiffusion technic of Johansson et al. (8) was used with Ionagar No. 2 (Oxoid) and 0.1% sodium azide as a preservative. Usually 1:10 and 1:50 dilutions of the antiserums gave optimal readings in the preliminary tests with measured amounts of the respective Ig class.

Lymph node extracts were prepared by sonification of tared amounts of the tissues followed by centrifugation at $2,000 \times g$. The supernate was concentrated to at least 1/20 of its original volume by dialysis against powdered baker's sugar and evaporation in vacuum. Dilution factors were considered in the evaluation of the tests. Controls with serums containing known amounts of the respective Ig class were included with each series of slides.

Antibody avidity

One to 1.3×10^6 borrelias per ml PBSCM were disintegrated in the Bronwill mill at 8,000 cycles for 10 min with CO_2 cooling and then centrifuged at $12,000 \times g$ and 2 to 3 C. The sediment was washed 3 times with an equal volume of PBSCM and recentrifuged. The supernates were pooled, evaporated with a fan, and concentrated by dialysis against powdered baker's sugar. Preliminary tests showed that optimal precipitation was obtained when 0.2 to 0.4 ml of this antigen were added to 1 ml Ig solution containing 10 mg/100 ml of the respective Ig (5). The precipitation was carried out for 2 hr at 37 C. Each sample was centrifuged at $12,000 \times g$ and the sediment washed three times with an equal volume of PBSCM. The final sediment was then divided into seven aliquots. Each of them was exposed to an equal volume of buffer with pH increasing stepwise by 0.5. Glycine HCl–glycine buffer was employed at pH 3.0 and 3.5, and acetate–acetic acid buffer at pH 4.0 to 6.0. Control tubes contained only the respective Ig antibody. The suspensions were shaken at room temperature for 30 min. Then the suspensions were chromatographed through a CM cellulose column (Whatman, Inc.) with 0.05 M sodium hydrocarbonate as an eluent. The effluents containing Ig were pooled and concentrated by pressure dialysis according to Tozer et al. (12). The amount of eluted Ig was determined by the method of Johansson and co-workers (8).

Hemagglutination test

The indirect hemagglutination test (IHA) was performed with serums and concentrated lymph node extracts, using the Takacsy appliances. Formolized sheep erythrocytes (Difco, Inc.) were tanned. Borrelias disintegrated in the Bronwill mill and adjusted to contain 10 mg protein/ml were adsorbed to the tanned cells for 60 min at 20 C. Twofold serial dilutions of the serums or lymph node extracts were prepared in PBSCM with 1% bovine albumin as a stabilizer. After mixing the reactants, the slides were rotated for 15 min, incubated for 2 hr at 37 C, and then overnight at 3 to 4 C. The highest antibody dilution giving a 3+ or 4+ agglutination was considered the endpoint.

The indirect hemagglutination test was performed also with isolated serum IgG, IgA and IgM but the relatively small amount of immuno-globulins in the lymph node extracts did not permit tests with individual Ig classes in all instances.

Immobilizine and borreliolysin determinations

Freshly collected live borrelias were suspended in PBSCM to contain 20 to 40 organisms per microscopic field at ×400 magnification.

In the immobilizine test, inactivated serum or lymph node extract was diluted twofold with PBSCM and added to an equal amount vol/vol of the borrelial suspension. After 30 min incubation the number of immobilized borrelias was counted in both the test slides and the controls that contained PBSCM but no antibody.

The borreliolysin test was carried out with the same *Borrelia* suspensions and serum dilutions. The serums, however, were not inactivated and 0.02 ml of an 1:100 dilution of guinea pig complement without preservative was added. The incubation time was 90 min at 37 C.

In both tests the highest dilution of serum, lymph node extract or Ig class causing immobilization and borreliolysis, respectively, was established and adjusted according to the number of motile and intact borrelias, respectively, in the control slides incubated without antibody.

Ribosomal fractions, RNA and lysine incorporation

Ribosomal fractions of lymph nodes were separated by differential centrifugation according to Talal (11).

RNA was extracted and estimated by the phenol–ether method with the addition of sodium lauryl sulfate and 1,5 naphthalene disulfonate by the procedure of Ramming and Pilch (9).

Total RNA in the entire lymph nodes was determined as well as the percent thereof appearing in the mitochondrial, ribosomal and postribosomal fractions.

Polypeptide synthesis was measured by the incorporation of L-^{14}C-lysine (New England Nuclear Lab.) according to Talal (11), using the postmitochondrial supernate as the source of synthetase. The results were expressed in counts per minute (cpm) as determined in the Beckman gas low-background counter. Puromycin 10^{-5} M was used as a control of the synthesis.

Borrelias in lymph node cells

Frozen sections of the lymph nodes were stained by the fluorescent antibody method, using rabbit anti-borrelia serum globulin prepared in this laboratory by the method of Uhr et al. (13). Sheep antirabbit globulin labeled with fluorescein isothiocyanate (Hyland Co.) was used in the indirect fluorescent antibody method. Carnoy-fixed tissue, stained by the Krajan and the Unna-Pappenheim methods, served as controls.

RESULTS

Median infective dose and pathogenicity in primates

The MID of *B. turicatae* for patas was 100 ± 12 organisms and 57 ± 7 for vervets. Borrelemia was observed between 3 to 10 days after infection. The mortality rate was $34 \pm 7\%$ in patas and $40 \pm 3\%$ in vervets.

The MID of *B. hermsii* was 73 ± 9 organisms in patas and 50 ± 7 in vervets. Borrelemia lasted from days 3 to 11 after infection. The mortality rate was $42 \pm 5\%$ in patas and $62 \pm 8\%$ in vervets. The results of the study of nonhuman primates used to establish these data are not included in the present communication.

Immunologically active cells, immunoglobulin and avidity estimations

Table 1 shows the geometric means and standard deviations of IgG, IgA and IgM levels, their avidity and the number of IAC.

The immunologically active cells in the lymph nodes increased after infection, reaching a peak on day 14. Vervets appeared to produce slightly more IAC than patas. The IAC formation appeared to be more intensive after *B. hermsii* than after *B. turicatae* infection but the large and inherent error in the IAC count makes statistical evaluation of the data difficult.

Serum IgG levels were higher in vervets than in patas monkeys on day 7 after infection. The serum IgG values reached the maximum on day 14 and were still elevated on day 30, whereas the number of cells producing this IgG class in the lymph nodes returned to values observed before infection on day 30. IgA levels increased after infection, remaining on an even level until day 30, with a small increase at day 14. IgM levels were elevated on day 7 but appeared to decrease in most instances by day 14 both in the serums and in the lymph nodes.

The avidity of antibody–antigen complexes is indicated by the pH necessary to split these complexes when the Tozer et al. method (12) is applied for the determination of avidity. Under these experimental conditions, pH 6.0 is considered the lowest pH splitting *Borrelia* – Ig bonds. Table 1 shows that the avidity of IgG increased strongly, whereas that of IgA and IgM increased in lesser magnitude. IgA and IgM capable of forming loose bonds with the borrelial antigens circulated on day 7, when borrelemia was present. Later formed IgG was more firmly attached to the antigen.

Serologic tests

Table 2 presents the results of the IHA, immobilizine and borreliolysin tests. All were negative (IHA < 1:8, immobilizine and lysin titers < 1:50) before infection. Vervets responded with somewhat higher titers than patas. *Borellia hermsii* elicited stronger reactions than *B. turicatae*. The titers of these reactions increased between days 14 and 30 in the serums and had reached significant levels in the lymph nodes already by day 7, during the borrelemia. The participation of IgM was greater than that of IgG in the early stages of infection; IgA participated in the IHA reaction, little in the immobilizine, and not at all in the borreliolysin test.

RNA determination and in vitro polypeptide synthesis

Table 3 shows that total RNA in the lymph nodes reached its maximum on day 7 after infection. The ribosomal fraction of RNA remained elevated on day 30, forming 31 ± 3 to 38 ± 3% of the total RNA. The RNA in the postribosomal supernate reached similar proportions on day 7 after infection but decreased later. This decrease was less marked after *B. hermsii* infection. The proportion of postribosomal RNA was higher than before infections. Polypeptide synthesis, indicated by the incorporation of L-^{14}C-lysine, was more active than before infection but declined on day 30. Puromycin, which inhibits polypeptide synthesis by interfering with the attachment of "soluble" RNA to ribosomes, blocked this reaction.

TABLE 1. Immunologically active cells, immunoglobulins, and their avidity[a]

Borrelia	Sample and primate		Assay	Days after infection[a]			
				-7	+7	+14	+30
turicatae	Serum	patas	IgG[b]/Av[c]	886 ± 33/N[d]	848 ± 31/4.3	1,034 ± 36/3.6	974 ± 30/3.1
		vervet		958 ± 41/N	1,005 ± 43/4.1	1,015 ± 34/3.8	984 ± 26/3.0
		patas	IgA/Av	176 ± 13/N	197 ± 18/5.4	166 ± 15/5.0	196 ± 16/4.3
		vervet		204 ± 16/N	180 ± 10/5.2	178 ± 15/5.0	190 ± 18/4.6
		patas	IgM/Av	214 ± 16/N	358 ± 21/5.6	318 ± 19/5.2	230 ± 15/4.3
		vervet		240 ± 20/N	332 ± 22/5.3	338 ± 23/5.0	262 ± 18/4.3
	Lymph node	patas	IAC[e]/IgG[f]	0[h]/0[h]	156 ± 18/20.2	280 ± 33/30.4	189 ± 20/25.4
		vervet		0/0	173 ± 19/22.4	303 ± 34/31.4	230 ± 21/36.6
		patas	IAC/IgA	0/0	23 ± 4/4.4	34 ± 5/4.1	22 ± 5/4.6
		vervet		0/0	33 ± 5/4.6	36 ± 4/4.8	28 ± 5/4.1
		patas	IAC/IgM	0/0	94 ± 8/17.5	93 ± 9/23.2	60 ± 9/13.2
		vervet		0/0	95 ± 9/15.4	107 ± 8/18.4	85 ± 9/10.5
hermsii	Serum	patas	IgG/Av	876 ± 30/N	890 ± 30/4.9	996 ± 33/3.8	922 ± 28/3.2
		vervet		938 ± 47/N	996 ± 41/4.9	995 ± 30/3.6	944 ± 27/3.5
		patas	IgA/Av	178 ± 15/N	198 ± 16/5.7	178 ± 17/5.4	183 ± 21/4.8
		vervet		214 ± 20/N	192 ± 19/5.5	182 ± 18/5.3	180 ± 17/4.6
		patas	IgM/Av	220 ± 19/N	366 ± 28/5.3	330 ± 23/5.0	232 ± 20/4.4
		vervet		232 ± 21/N	340 ± 26/5.6	342 ± 26/5.0	244 ± 21/4.6
	Lymph node	patas	IAC/IgG	0/0	216 ± 23/21.7	311 ± 35/32.5	195 ± 21/26.3
		vervet		0/0	284 ± 31/31.5	315 ± 29/31.9	199 ± 25/30.1
		patas	IAC/IgA	0/0	26 ± 4/4.6	36 ± 3/4.2	20 ± 3/4.4
		vervet		0/0	31 ± 4/4.4	34 ± 4/4.6	25 ± 6/4.2
		patas	IAC/IgM	0/0	105 ± 9/20.3	115 ± 11/26.4	66 ± 10/12.9
		vervet		0/0	110 ± 10/16.0	122 ± 12/20.1	73 ± 10/9.7

[a] Day of infection: 0. [b] In mg/100 ml, mean ± SD. [c] Avidity expressed as the mean of the highest pH of the buffers required to dissociate the antigen–antibody complexes. [d] No antigen–antibody complex was formed. [e] Number of cells showing fluorescence with antiserum against the respective Ig class, per 10^8 cells. [f] In mg per g tissue, mean. [g] Mean less than 10 Ig producing cells per 10^8 cells. [h] Mean less than 1 mg per g tissue.

TABLE 2. Indirect hemagglutination, immobilizine, and borreliolysin tests

Sample and primate		Component	Test	Day after infection[a]		
				7	14	30
Serum	patas	T[b]	IHA[c]	8 ± 5/10 ± 3[d]	34 ± 8/54 ± 4	104 ± 7/210 ± 41
	vervet			18 ± 8/24 ± 10	54 ± 10/61 ± 9	164 ± 28/280 ± 38
	patas	IgG		0/0[e]	8 ± 3/10 ± 4	62 ± 9/114 ± 20
	vervet			0/8 ± 2	12 ± 4/22 ± 4	78 ± 8/132 ± 21
	patas	IgA		0/0	10 ± 3/8 ± 4	10 ± 3/10 ± 7
	vervet			0/0	8 ± 4/11 ± 4	16 ± 3/18 ± 4
	patas	IgM		0/0 ± 4	18 ± 5/30 ± 5	38 ± 7/105 ± 21
	vervet			14 ± 4/10 ± 4	32 ± 8/26 ± 4	76 ± 8/132 ± 16
Lymph	patas	T		16 ± 3/16 ± 5	40 ± 4/28 ± 3	42 ± 7/84 ± 6
node	vervet			30 ± 8/34 ± 7	68 ± 7/89 ± 9	68 ± 6/92 ± 10
Serum	patas	T	Imm[f]	0/0[g]	58 ± 10/62 ± 8	246 ± 29/272 ± 32
	vervet			0/0	64 ± 12/83 ± 15	315 ± 32/339 ± 37
	patas	IgG		0/0	0/0	85 ± 15/82 ± 21
	vervet			0/0	0/0	96 ± 19/107 ± 22
	patas	IgA		0/0	0/0	0/51 ± 8
	vervet			0/0	0/0	51 ± 12/55 ± 10
	patas	IgM		0/0	51 ± 7/52 ± 9	220 ± 22/158 ± 19
	vervet			0/0	53 ± 9/57 ± 8	170 ± 21/171 ± 18
Lymph	patas	T		51 ± 8/52 ± 7	64 ± 8/69 ± 9	123 ± 20/144 ± 17
node	vervet			53 ± 9/55 ± 8	61 ± 9/63 ± 10	128 ± 18/172 ± 19
Serum	patas	T	Lys[h]	0/θ[g]	0/0	124 ± 21/150 ± 19
	vervet			0/0	0/0	148 ± 18/162 ± 17
	patas	IgG		0/0	0/0	60 ± 11/78 ± 16
	vervet			0/0	0/0	80 ± 12/78 ± 15
	patas	IgA		0/0	0/0	0/0
	vervet			0/0	0/0	0/0
	patas	IgM		0/0	0/0	72 ± 13/70 ± 14
	vervet			0/0	0/0	75 ± 10/81 ± 12
Lymph	patas	T		53 ± 7/51 ± 5	71 ± 8/78 ± 9	66 ± 5/65 ± 12
node	vervet			55 ± 8/57 ± 6	76 ± 8/79 ± 10	61 ± 9/68 ± 10

[a] Day of infection: 0. All tests were negative on day −7.   [b] Total serum or lymph node extract.   [c] Indirect hemagglutination test, reciprocal titers, mean ± SD.   [d] In primates infected with *B. turicatae*/in primates infected with *B. hermsii*.   [e] Mean less than 8.   [f] Immobilizine test, reciprocal titers.   [g] Mean less than 50.   [h] Borreliolysin test, reciprocal titers.

Borrelias in lymph nodes

Borrelias were found in the lymph nodes on day 7 in all infected animals and on day 14 in 6 of the 30 infected primates (20%). They were never found inside cells. On day 14 fragments of borrelias could be seen inside macrophages in the lymph nodes of 28 of the 30 primates. Technical difficulties involved in the counting of the potentially phagocytic groups of cells precluded the calculation of the ratio of total potential phagocytes and those containing fragments of borrelias and antigen–antibody complexes. After 30 days no fragments could be distinguished by the fluorescent antibody method. An attempt to label borrelias with a radioisotope and to enumerate simultaneously phagocytic cells is the sub-

TABLE 3. Lymph node RNA and ribosomal polypeptide synthesis

Borrelia	Assay	Primate	Day after infection[a]		
			−7	+7	+30
turicatae	Total RNA	patas	4.7 ± 0.5[b]	7.6 ± 1.0	5.5 ± 0.7
		vervet	5.1 ± 0.4	9.1 ± 1.1	6.3 ± 0.8
	Ribosomal RNA	patas	22 ± 3[c]	28 ± 3	31 ± 3
		vervet	26 ± 4	31 ± 4	33 ± 5
	Postribosomal RNA	patas	15 ± 3[d]	28 ± 4	19 ± 4
		vervet	19 ± 4	32 ± 5	21 ± 4
	L-¹⁴C-lysine inc.[e]	patas	215 ± 17[f]	453 ± 43	270 ± 20
		vervet	230 ± 22	487 ± 41	295 ± 27
	L-¹⁴C-lysine + puromycin[g]	patas	31 ± 4	36 ± 7	35 ± 6
		vervet	29 ± 5	38 ± 5	33 ± 4
hermsii	Total RNA	patas	4.8 ± 0.6	7.7 ± 0.9	5.3 ± 0.5
		vervet	5.3 ± 0.6	9.9 ± 1.0	5.8 ± 0.7
	Ribosomal RNA	patas	21 ± 4	31 ± 5	38 ± 3
		vervet	25 ± 3	33 ± 4	35 ± 5
	Postribosomal RNA	patas	14 ± 3	30 ± 4	24 ± 4
		vervet	18 ± 3	33 ± 3	25 ± 4
	L-¹⁴C-lysine inc.	patas	220 ± 19	483 ± 46	291 ± 31
		vervet	238 ± 20	499 ± 51	307 ± 28
	L-¹⁴C-lysine + puromycin	patas	29 ± 3	34 ± 4	35 ± 3
		vervet	32 ± 4	33 ± 4	34 ± 3

[a] Day of infection: 0.   [b] In mg per g tissue, mean ± SD.   [c] % of total recovered RNA in the ribosomal fraction.   [d] % of total recovered RNA in the postribosomal fraction supernate.   [e] L-¹⁴C-lysine incorporated.   [f] Counts per minute (cpm) per mg of polyribosomal protein, mean ± SD.   [g] Complete L-¹⁴C-lysine incorporating mixture plus 10^{-5} M puromycin, cpm per mg ribosomal protein, mean ± SD.

ject of a preliminary publication (submitted).

DISCUSSION AND CONCLUSIONS

Borreliosis in primates is a cyclic disease caused by a cyclic agent. We have been fortunate to obtain two strains with minimal cyclic variations. This enabled us to investigate some features of the immunologic pattern in borreliosis.

It appeared that borrelias multiply in the blood and in the lymphatic tissue. The initial response was the appearance of immunologically active cells which first formed a greater amount of IgM class antibodies of low avidity followed by the appearance of class IgG antibodies possessing higher avidity but a smaller num-

ber of binding sites. Hence there is a greater participation of IgM in the serologic tests. IgA did not seem to play a significant role in borreliosis, perhaps because of its inability to fix complement. IgD apparently does not participate in the immunologic process when studied by routine methods (5). Phagocytes did not seem to attack live borrelias (3, 5). This was indicated also in the present experiments. They function perhaps only as scavengers whereas the principal cause of the destruction of borrelias in circulation appeared to be antibodies formed by the immunologically active cells. The increased production of protein and RNA apparently indicated a stimulation of globulin formation induced by borrelias or their products. This has been observed also in a preliminary study (4).

The late (60 and more days after infection) phases of the immunologic picture are still under study. No antigen could be found in the lymph nodes 30 days after infection. Therefore the stimulus for late antibody formation requires further investigation.

Unquestionably more attention will have to be devoted to the role of the spleen and the liver. Data available to date, based on the study of open abdominal biopsies, are insufficient to draw valid conclusions. However it was shown in previous studies that the histologic picture and the immunologic functions of the spleen closely parallel those of the superficial lymph nodes (4).

Therefore it may be concluded that the defense mechanism of the patas and vervet is primarily based on cell-bound antibodies formed in the lymphatic system. When these products are released into the circulation in sufficient quantities, disintegration of the blood-stream borrelias ensues. It is probable that borrelias in organs are killed in the same manner. Phagocytes appear to play a secondary role by taking up antigen–antibody complexes and fragments of borrelias. One theory presently accepted by many investigators indicates that so-called macrophages mediate the impulse for the formation of immunologically active globulin by immunocompetent cells. However, the chemistry of this process has not yet been explored. Therefore further investigation of this problem is strongly indicated, particularly in chronic and recurrent diseases, because the question of memory cells is still not fully elucidated.

REFERENCES

1. CRABBÉ, P. A., AND J. F. HEREMANS. Gastroenterology 51: 305, 1966.
2. FELSENFELD, O. Borrelia. St. Louis, MO: Green, 1971, p. 74.
3. FELSENFELD, O. Borrelia. St. Louis, MO: Green, 1971, p. 94.
4. FELSENFELD, O., H. B. GOLDSTEIN, R. H. WOLF AND J. HOLMES. Exptl. Molec. Pathol. 12: 225, 1970.
5. FELSENFELD, O., AND R. H. WOLF. Acta Tropica 26: 156, 1969.
6. FELSENFELD, O., AND R. H. WOLF. Ann. Trop. Med. Parasitol. 67: 335, 1973.
7. FINE, J. M., AND M. STEINBUCH. Rev. Europ. Etud. Clin. Biol. 15: 1115, 1970.
8. JOHANSSON, S. G. O., C. F. HÖGMAN AND J. KILLANDER. Acta Pathol. Microbiol. Scand. 74: 519, 1968.
9. RAMMING, K. P., AND Y. H. PILCH. Transplantation 7: 286, 1969.
10. REIF, A. E. Immunochemistry 6: 723, 1969.
11. TALAL, N. J. Biol. Chem. 241: 2067, 1966.
12. TOZER, B. T., K. A. CAMMACK AND H. SMITH. Biochem. J. 84: 80, 1962.
13. UHR, J. W., S. B. SALVIN AND A. M. PAPPENHEIMER, JR. J. Exptl. Med. 105: 11, 1957.

Teratogenic effects of triamcinolone on the skeletal and lymphoid systems in nonhuman primates[1]

A. G. HENDRICKX, R. H. SAWYER, T. G. TERRELL,
B. I. OSBURN, R. V. HENRICKSON, AND A. J. STEFFEK[2]

California Primate Research Center
University of California, Davis, California 95616

Corticosteroids are used to treat several human diseases, especially allergies, and inflammatory, dermatologic and salt-imbalance conditions. Because of the possibility that treatment for one of these conditions might occur before a mother knows she is pregnant and that an effect may be exerted over a period of time which could encompass the organogenic or sensitive period, it is important to establish the teratogenic effect of these drugs. Walker (10, 11) has demonstrated that corticosteroids vary considerably in their teratogenic effects. Of the six corticosteroids tested he found triamcinolone to be one of the most teratogenic. Furthermore, Walker (11) observed that the dosage of triamcinolone required to produce a specific defect, cleft palate, was as different between breeds of rabbits as it was between rabbits and mice.

The initial objectives of the present study were to compare the teratogenic effects of this drug in three species of nonhuman primates and to develop an animal model for the study of orocraniofacial defects. Observations in the initial experiments indicated that the thymus was affected at a high frequency in the rhesus and bonnet monkeys; consequently, the experiments were extended to include thymus differentiation during later stages of pregnancy and to determine whether the effects of triamcinolone on the fetus result in permanent teratological effects on the lymphoid system.

MATERIALS AND METHODS

Husbandry and breeding

Sexually mature female bonnet monkeys (*Macaca radiata*), rhesus monkeys (*M. mulatta*), and baboons (*Papio cynocephalus*) were housed individually in aluminum cages and maintained in compliance with the

[1] This study was supported by Public Health Service National Institutes of Health Grants RR00169, GMS00537, and DEDH03927-01.
[2] Address: American Dental Association, Chicago, Illinois.

standards of the Federal Animal Welfare Act and Institute for Laboratory Animal Resources (ILAR). Isoniazid, at a dosage of 10 mg/kg per day per animal, was incorporated into the diet as a prophylactic agent against tuberculosis. Menstruation was detected by daily visual examination of vaginal swabs. All females were bred once, 2 days before midcycle. Midcycle was calculated individually for females and was based on their previous menstrual cycles. Mating was limited to 2 hours and the day of mating was designated as day 0 of pregnancy. Pregnancy was confirmed by bimanual palpation of the uterus per rectum.

Dosing

Triamcinolone acetonide was administered intramuscularly to the mother for 1 or 4 consecutive days. The initial doses to all three species were either 25 or 75 mg/day, but because of the variation in the body weight of individual animals, subsequent doses were calculated on a body weight basis. Five bonnet monkeys received 15–20 mg/kg per day of triamcinolone on days 41–44 of pregnancy (Table 1), 10 rhesus monkeys received 3–28 mg/kg per day between days 37–48 (Table 2), and 6 baboons received 1–14 mg/kg per day between days 37–44 (Table 3). In addition, 10 rhesus monkeys received 3–19 mg/kg per day between days 50–133 of pregnancy (Table 4). Ten untreated control fetuses of each species were also used in this study.

Recovery and examination of fetuses

The fetuses were delivered by cesarean section at 100 days of gestation, or near term, except several cases which were allowed to deliver naturally (Tables 1–4). The fetuses were immediately weighed, measured and photographed. After evisceration, the skeleton was stained with

TABLE 1. Teratogenicity of triamcinolone in bonnet monkeys

Animal no.	Days of pregnancy	Dose[a]	Pregnancy outcome and condition of fetus/infant
D-4	41–44	15	Cesarean delivery at 157 days gestation, died 1 day postpartum; cleft palate, craniofacial dysmorphia, hyperextension of left knee and medial rotation of limb, hypoplastic thymus, growth retardation
E-8	41–44	15	Stillborn at 178 days gestation, craniofacial dysmorphia, hyperextension of left knee and medial rotation of limb, cutaneous webbing of all digits, hypoplastic thymus and adrenal, aplastic renal pelvis
E-4	41–44	15	Cesarean delivery at 160 days gestation, dead in utero, craniofacial dysmorphia, hypoplastic thymus, growth retardation
D-7	41–44	18	Cesarean delivery at 100 days gestation, dead in utero, craniofacial dysmorphia, hypoplastic thymus, growth retardation
D-3	41–44	20	Cesarean delivery at 100 days gestation, viable, craniofacial dysmorphia, hypoplastic thymus, growth retardation

[a] In mg/kg per day.

TABLE 2. Teratogenicity of triamcinolone in rhesus monkeys

Animal no.	Days of pregnancy	Dose[a]	Pregnancy outcome and condition of fetus/infant
M-3	38	28	Cesarean delivery at 183 days gestation, died 2 days postpartum; craniofacial dysmorphia, syndactyly, hypoplastic phalanx and cutaneous webbing of hindfeet, hypoplastic thymus, lymphoid hypoplasia
M-6	37–40	5	Abortion at 115 days gestation; no detectable defects
L-6	41–44	3	Cesarean delivery at 116 days gestation; resorption, empty endometrial sac recovered
M-5	41–44	5	Cesarean delivery at 100 days gestation, viable; no detectable defects
N-2	41–44	5	Cesarean delivery at 160 days, grossly normal at birth; sacrificed at 74 days of age; hypoplastic thymus, function of lower limbs impaired
N-7	41–44	11	Cesarean delivery at 83 days gestation, dead in utero—no evaluation
	41–44	15	Resorption, no tissue recovered
N-6	41–44	25	Cesarean delivery at 143 days gestation, dead in utero—cord extrangulation, hypoplastic thymus, growth retardation
M-2	42–45	13	Cesarean delivery at 161 days gestation, dead in utero; craniofacial dysmorphia, hypoplastic thymus and adrenals, growth retardation
M-8	45–48	12	Cesarean delivery at 100 days gestation, viable; craniofacial dysmorphia, hypoplastic thymus, growth retardation

[a] In mg/kg per day.

alizarin red S and the viscera were fixed in 10% buffered formalin. The tissues were processed in a routine manner, sectioned at 6 μm, and stained with hematoxylin and eosin.

Lymphocyte physiology

Lymphocyte purification: Peripheral blood lymphocytes were obtained by centrifugation on a Ficoll-Hypaque gradient. Lymphocytes were washed three times with Earl's minimum essential media and adjusted to a final concentration of 4×10^6 cells/ml in the medium.

Detection of erythrocyte rosette-forming lymphocytes: The technique for form-ing erythrocyte rosettes as described by Jondal et al. (8) was followed with slight variation. Fetal calf serum absorbed with sheep erythrocytes and heat inactivated at 56 C for 60 min was added to the mixed cell suspension for a final concentration of 10%. The suspension was incubated at 4 C overnight following the final centrifugation. One drop of 0.2% trypan blue in phosphate buffered saline was added to the cell suspension immediately prior to resuspension for detection of nonviable cells.

Detection of immunoglobulin on lymphocytes: Lymphocytes (2×10^6) were incubated with either (rabbit) anti-human IgM serum (Miles Laboratories, Kankakee, Ill.) or (rabbit) anti-monkey IgG serum (Cappel Laboratories, Downington, Pa.) for 30 min at 37 C. Both antisera were conjugated with fluorescein isothiocyanate. The cells were then washed twice with phosphate buffered saline containing 0.1% sodium azide. The cells were suspended in phosphate buffered glycerol, and the number of fluorescing cells per 100 lymphocytes was counted.

RESULTS

General

The effects of triamcinolone during the embryonic period (18–47 days gestation) consisted of resorption, intrauterine death, and malformations. Defects were seen in the orocraniofacial region, thorax and hindlimbs, and the thymus, adrenal and kidney. Changes in the hair coat and skin were also observed but will not be considered here. Resorption was observed only in the rhesus monkey, but intrauterine death was observed in all three species. The most severe defects of the orofacial region consisted of cleft palate in a bonnet monkey (Fig. 1), and choanal atresia and mandibular overbite in a baboon. Other skeletal defects consisted of two cases of hyperextension of the knee on the left side and medial rotation of the hindlimbs in bonnet monkeys, and depression of the sternum toward the spine (funnel chest) in a baboon. Syndactyly, characterized by the fusion of the distal phalanges of the second and third digits, hypoplasia of the fifth digit of the feet (Fig. 2), and cutaneous webbing between the second, third and fourth digits, was observed in a rhesus monkey. Three bonnet monkeys and eight rhesus monkeys also had alterations in facial development characterized by forward protrusion of the forehead, widening of the head, and expansion of the cranial sutures and both the frontoparietal

TABLE 3. Teratogenicity of triamcinolone in baboons

| Animal no. | Treatment | | Pregnancy outcome and condition of fetus/infant |
	Days of pregnancy	Dose[a]	
14	37–40	1.0	Abortion at 80 days gestation, no detectable defects
15	37–40	1.0	Cesarean delivery at 170 days of gestation, present age 2.4 years, growth retardation, lymphocyte deficiency
12	41–44	1.6	Natural birth at 182 days gestation, present age 3.4 years, choanal atresia, mandibular overbite, funnel chest, growth retardation, lymphocyte deficiency
10	41–44	1.0	Cesarean delivery at 111 days gestation, viable, no detectable defects
11	41–44	1.6	Cesarean delivery at 101 days gestation, viable, no detectable defects
13	41–44	14.0	Maternal death due to toxicity, in utero death at 76 days gestation, no detectable defects

[a] In mg/kg per day.

TABLE 4. Teratogenicity of triamcinolone in fetal rhesus monkeys

Animal no.	Days of pregnancy	Dose[a]	Pregnancy outcome and condition of fetus/infant
	Treatment		
R-3	50–53	13	Cesarean delivery at 100 days gestation, viable; craniofacial dysmorphia, hypoplastic thymus, growth retardation
R-9	60–63	13	Cesarean delivery at 101 days gestation, viable; craniofacial dysmorphia, hypoplastic thymus, growth retardation
S-1	61–64	14	Cesarean delivery at 160 days gestation, died 2 days postpartum; cysts on liver, growth retardation
S-3	62–65	19	Natural delivery at 143 days gestation, died shortly after birth; craniofacial dysmorphia, hypoplastic thymus, hypoplastic adrenals
R-5	65–68	14	Cesarean delivery at 100 days gestation, viable; craniofacial dysmorphia, hypoplastic thymus, growth retardation
S-7	70–73	3	Cesarean delivery at 113 days gestation, viable; hypoplastic thymus
R-1	70–73	13	Cesarean delivery at 102 days gestation, viable; craniofacial dysmorphia, hypoplastic thymus, growth retardation
T-4	100–103	13	Cesarean delivery at 163 days gestation, died in utero; hypoplastic thymus, growth retardation
T-5	113–116	9	Natural delivery at 203 days gestation, dead in utero (at 175 days) —no evaluation
T-3	130–133	9	Cesarean delivery at 178 days gestation, dead in utero; hypoplastic thymus

[a] In mg/kg per day.

and occipitoparietal fontanelles. This condition becomes more pronounced with the age of the fetus and the general appearance is that of a slight to moderate depression at the bridge of the nose.

Malformations of the viscera, with the exception of the thymus, were infrequent and showed no specific pattern. Adrenal aplasia and hypoplasia occurred in bonnet and rhesus monkey fetuses respectively, and cysts were observed on the liver in a rhesus monkey (Tables 1, 2). Malformed kidneys were also observed once in a bonnet monkey. The malformations consisted of absence of the pelvis with a heavy thickening of the renal capsule.

Growth retardation was common to all three species but was most marked in bonnet monkeys. The decrease in body size (weight and length) was 30–40% of comparably aged controls.

Thymus

The thymus was the most severely affected of all the organs. Grossly, it was hypoplastic in all the bonnet monkeys and in 13 of 17 rhesus monkeys examined at necropsy (Tables 2 and 4), but appeared to be normal in the four baboon fetuses.

The histologic appearance of the normal 100-day fetal thymus from the rhesus monkey is shown in Fig. 3. It is differentiated into cortical and medullary zones with the usual complement of lymphoid and epithelial elements in each region. Exposure of the dam to triamcinolone between 45–70 days gestation resulted in a marked hypoplasia of the thymus in fetuses sacrificed at 100 days gestation (Fig. 3). There is almost a total depletion of thymic lymphocytes and a reduction in the epithelial component. The lobular structure of the thymus is still present. In addition, a depletion of lymphocytes from the thymic dependent areas of the spleen and lymph nodes was observed, and will be presented in another publication.

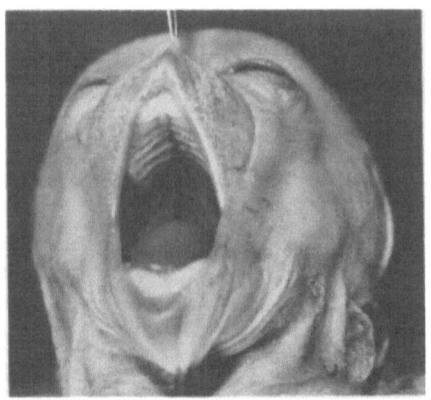

Figure 1. Newborn bonnet monkey with median cleft of the secondary palate. The cleft extended the length of the maxilla but did not involve the premaxilla. Treatment with triamcinolone (15 mg/kg per day) was on days 41–44 of pregnancy.

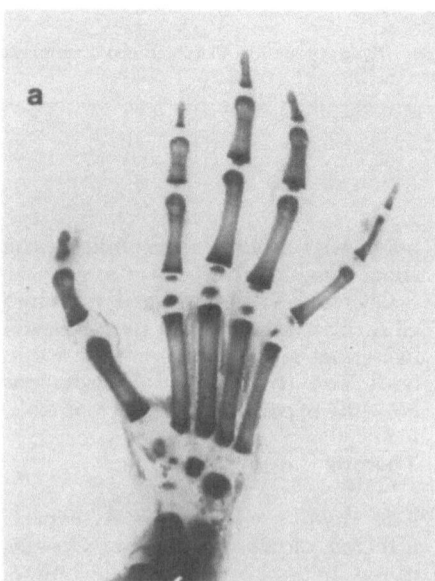
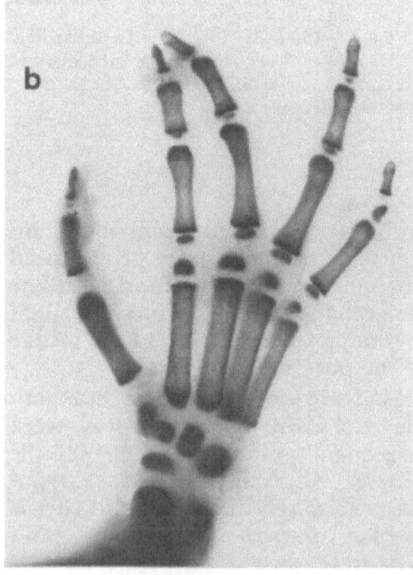

Figure 2. Right feet of newborn rhesus monkeys. *a*) Normal, untreated control, *b*) Abnormal triamcinolone treated (28 mg/kg) on day 38 of pregnancy. The distal phalanges of the second and third digits were fused and the medial phalanx of the fifth digit was hypoplastic. Both specimens were stained with alizarin red S.

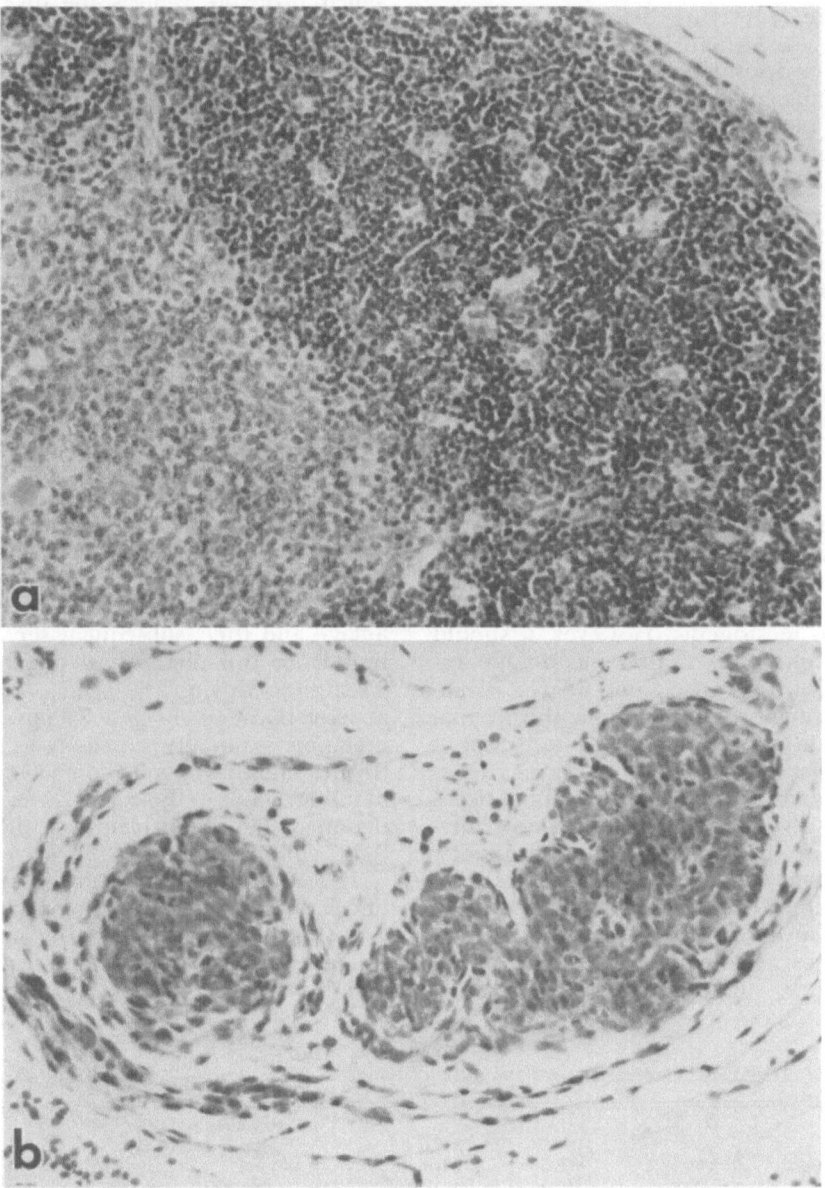

Figure 3. *a*) A normal thymus from a 100-day rhesus monkey fetus. It is well differentiated into cortical and medullary regions. ×236. *b*) A hypoplastic thymus from a 100-day rhesus monkey fetus that had been exposed to triamcinolone (13 mg/kg per day) on days 60–63 of development. The cortical and medullary regions are indistinct and the organ is depleted of lymphocytes. ×236.

Lymphocytes

The surviving baboon offspring appeared deficient in both total and subpopulations of lymphocytes (Table 5). Both animals demonstrated a persistent lymphopenia with a marked decrease, as compared to healthy adult baboons, in the number of peripheral blood lymphocytes which form spontaneous nonimmune rosettes with sheep erythrocytes (erythrocyte rosettes). An increased percent of the peripheral blood lymphocytes in triamcinolone treated baboons had detectable surface immunoglobulin (Table 5).

DISCUSSION

The effect of administration of triamcinolone during the embryonic period in bonnet monkeys, rhesus monkeys, and baboons indicates that it is teratogenic. However, it should be emphasized that the dosage required to cause some degree of abnormality is 3–28 times the human therapeutic dosage, on a body weight basis, in both bonnet and rhesus monkeys, while malformations were observed in baboons at levels equivalent to the human therapeutic dosage. The fact that the effects range from resorption to normal offspring after treatment with similar doses at similar times during pregnancy suggests considerable vari-

ability in response of these species, which probably reflects the diverse genotypes common to nonhuman primate populations. The variation in observed malformations suggests that either this drug interferes with several developmental pathways or that the time of treatment occurred either before or after the maximum teratogenic sensitive period for a given structure. As a result, minor defects are more frequently seen.

The large number of both bonnet and rhesus monkeys with hypoplastic thymuses suggests that triamcinolone acts in a specific way on this organ. The teratogenic sensitive period cannot be specifically determined from this study, but the thymus is sensitive to some degree between days 38 and 133 of pregnancy in the rhesus monkey. The early treatments, days 37–48, coincide with early proliferation of the thymic anlage and their subsequent separation from the pharyngeal pouches. Treatment during the fetal period, at least between 50 and 73 days of gestation, is equally damaging to the thymus and may involve some degree of interference in the actual differentiation of lymphoid elements within the thymic rudiment.

The impairment of normal thymus development by triamcinolone observed in this study suggests some similarities to the condition observed in mice (4) and humans (5). In all

TABLE 5. Lymphocyte studies on the offspring of triamcinolone treated baboons

Animal no.	No. of Trials	WBC × 10^3/cm³	Total lymph × 10^3/cm³	% Erythrocyte rosette forming cells	% IgG[a] bearing cells	% IgM[a] bearing cells
Pan 12	2	6.75	4.5	7.0	32.0	8.4
Pan 15	2	4.75	3.3	4.5	15.6	17.0
Control	2	10.63	7.6	32.5	15.2	2.2

[a] Single trial.

instances the impairment is related to abnormal development of the embryonic thymus.

In general, fetal death increased in bonnet and rhesus monkeys treated during the later stages of pregnancy. Furthermore, treatment after 100 days gestation consistently resulted in fetal death. The influence of corticosteroids on fetal maturation should not be overlooked. The relatively high incidence of growth retardation in both grossly normal and abnormal fetuses and infants indicates an inhibitory effect on the maturation process. Corticosteroids have been shown to impair placental development (1) and to be responsible for low birth weights (9).

It is apparent that treatment of pregnant animals with triamcinolone, in some cases with doses comparable to those recommended for humans, causes pronounced and permanent teratological effects on the lymphoid system of the offspring. The markedly reduced total white cells, lymphocytes, and erythrocyte-rosetting cells, and the increased numbers of cells with immunoglobulins on the surface, strongly suggest that the most profound effects of triamcinolone are on the thymic derived or T-lymphocyte population (7, 8, 12, 14). Similar observations relative to the effects of corticosteroids on the T-lymphocyte populations have been reported in a number of different species (2). In contrast to postnatal individuals, the treated fetal baboon is apparently incapable of complete reconstitution of lymphoid populations. This may not be the case in prenatally treated rats (3).

Corticosteroids are known to effectively reduce peripheral blood lymphocytes. Also, the wide use of these substances as immunosuppressive and anti-inflammatory agents has often been associated with their effects on T-lymphocytes. Investigations currently under way are designed to determine if offspring of triamcinolone treated pregnant animals have a permanent suppressive effect on T-lymphocyte mediated immune responsiveness.

A possible explanation of the difference in dose response to triamcinolone in the baboon compared to the bonnet and rhesus monkeys was not determined in this study. The possibility that the difference exists at the metabolic level should be considered because all three species demonstrate a close similarity in both embryonic development and in thalidomide teratogenicity (6).

Although triamcinolone has not been implicated as a human teratogen, cortisone, a closely related corticosteroid, was a suspected teratogen until recently, when a thorough review of epidemiological data indicated that it probably was not harmful (13). The present contraindications regarding the use of this drug include pregnancy, particularly during the first trimester. The results of the research reported here indicate that the warning regarding use should be extended through the entire period of pregnancy because of the potentially harmful and irreversible effect of this drug on the immune system.

REFERENCES

1. BLACKBURN, W. R., H. S. KAPLAN AND D. S. McKAY. *Am. J. Obstet. Gynecol.* 92: 234–246, 1965.
2. CLAMAN, H. N. *N. Engl. J. Med.* 278: 388–397, 1972.
3. FEIN, A., A. ORNOY AND L. NEBEL. *J. Anat.* 117: 223–237, 1974.
4. FLANNAGEN, S. P., *Genet. Res.* 8: 295–309, 1966.
5. GOOD, R. A. *Hosp. Prac.* 2: 38–53, 1967.

6. HENDRICKX, A. G., R. H. SAWYER, B. L. LASLEY AND R. D. BARNES. *Lab. Anim. Handb.* 6: 305–315, 1975.

7. JOHANSEN, K. S., T. S. JOHANSEN AND D. W. TALMAGE. *J. Allergy Clin. Immunol.* 54: 86–93, 1974.

8. JONDAL, M., G. HOLM AND H. WIGZELL. *J. Exptl. Med.* 136: 207–215, 1972.

9. NEBEL, L., AND A. ORNOY. *Advan. Exptl. Med. Biol.* 27: 250–256, 1972.

10. WALKER, B. E. *Science* 149: 862–863, 1965.

11. WALKER, B. E. *Proc. Soc. Exptl. Biol. Med.* 125: 1281–1284, 1967.

12. WALLEN, W. C., R. H. NEUBAUER, H. RABIN AND J. L. CICMANEC. *J. Natl. Cancer Inst.* 51: 967–975, 1973.

13. WILSON, J. G. *Environment and Birth Defects.* New York and London: Academic, 1973, p. 305.

14. WYLRAN, J., M. C. CARR AND H. H. FUDENBERG. *J. Clin. Invest.* 51: 2537–2543, 1972.

Tetrahydrocannabinol-induced manifestations of the "marihuana syndrome" in group-living macaques[1]

E. N. SASSENRATH AND L. F. CHAPMAN

*California Primate Research Center and Department of
Behavioral Biology, University of California School of Medicine
Davis, California 95616*

The issue of whether long-term frequent use of marihuana can lead to marked persistent changes in behavior and neuroendocrine function in man is still being debated. While the immediate short-term behavioral effects of the drug can be readily measured in a laboratory setting, the characterization of slowly-developing drug-induced changes in individually characteristic aspects of emotional and social behavior is more difficult. The nature of sample selection and the uncontrolled environmental variables in studies of chronic marihuana users versus nonusers have made the conclusions of such studies open to question. At the same time, the relevance of well-controlled rodent studies to the situation in man is questionable due to species differences in behavior and physiology. The present study of effects of long-term administration of Δ^9-tetrahydrocannabinol (THC), the principal active component of marihuana, to selected members of group-living macaque monkeys was established to bridge this gap. In particular, it was de-signed to utilize a pure drug form (as opposed to potentially variable cannabis extracts) in a primate social group test system under conditions of daily drug administration at levels comparable to heavy marihuana use in man.

In its conception and development the study has been oriented toward the detection of changes in affect or social behavior and correlated endocrine systems which could help to resolve contradictory and controversial claims for effects of long-term frequent use of cannabis—ranging from an "amotivational syndrome" among chronic North American users (6) to enhanced mental acuity and physical work potential among Jamaican users at much higher levels of daily drug intake (1). In addition, the long-term THC-treated monkeys of the present study—in comparison with their un-

[1] This work was supported by Public Health Service grants DA00135, RR00169, and MH21366.

drugged cagemates—have become appropriate subjects for investigations related to recent reports of testosterone depression and cerebral atrophy in heavy users of marihuana compared to nonusers (2, 5).

The results of the study to date have confirmed the absence of pronounced behavioral or endocrine derangements in the long-term THC-exposed subjects. However, a consistent persistent change in affect or "personality" has been noted which is manifest in a social context as increased irritable aggressiveness. The data further suggest that this behavioral change becomes more evident after the development of behavioral tolerance to the initial effects of THC which are characteristic of the marihuana "high."

SUBJECTS AND PROCEDURES

Subjects were pubertal or young adult macaques (*Macaca mulatta* or *M. fascicularis* (irus) born at the California Primate Research Center; all had been caged in stable social groups of 3 to 6 members for periods of 2 months to 2 years prior to initiation of the drug study. Social groups included six 3- to 4-membered "high stress" groups, caged under high density conditions to induce a wider range of chronic stress levels between dominants and subordinates, and two 6-membered peer groups within a 2-month age range, caged in larger outdoor facilities. Prior to the start of drugging, individual social roles within the dominance structure of each group were well characterized by repeated detailed observations to establish individual behavioral profiles for each member of each group. In Table 1, the composition of all cage groups is shown. As can be seen, the one or two drug-treated members in each group were selected to give an overall sampling of both males and females and dominant and sub-

ordinate group members. THC[2] was administered orally on preferred food at 2.4 mg per kg to drug-treated subjects once daily at one of three specified times, i.e., 8:30 AM, 12:00 N, or 3:30 PM. The daily oral dose of 2.4 mg/kg THC[3] was selected from preliminary dose-response studies as the maximum chronic level producing clearly detectable behavioral effects in monkeys during short-term chronic administration, while still permitting normal motor function and arousal in response to environmental stimuli. It thus did not render the monkey physically incapable of normal social interaction.

Observations of group social behavior were made on a continuing basis throughout the study utilizing a repertoire of over 50 behavioral items in the categories of aggression, affiliation, play, sexual behavior, neutral interactions, noninteractive behaviors, and stereotypy. Each cage group was observed for a total of 24 hours during the 2-month predrug baseline period and during four 2-month intervals spaced throughout the yearlong course of daily THC administration. Daily observations included sampling of both morning and afternoon spontaneous group interactions and stimulated aggressive interaction in a food competition situation. Reactants and behaviors were recorded in a coded shorthand notation in a computer-compatible format, and data were subsequently processed by computer to give total frequencies per session of all dyadic interactions and noninteractive be-

---

[2] Δ[9]-tetrahydrocannabinol (THC) was supplied by the National Institute on Drug Abuse.

[3] It can be estimated that this dose is equivalent to a daily intake of 0.4 mg/kg of THC by smoking in man, using a factor of ⅓ to correct for differences in weight per body surface area and a factor of ½ to relate peak responses by oral versus inhalation routes.

TABLE 1. Social group environments of long-term THC-treated subjects

Cage group	Macaque species[a]	Rank and sex	Age[b]	Caging	THC-treated
OE	M.mu.	#1,2,3 M #4,5,6 F	3-0 to 3-2	Outdoor: enriched	#2M #4F
OW	M.mu.	#1,2,3 M #4,5,6 F	3-0 to 3-2	Outdoor: enriched	#1M #5F
F	M.mu.	#1,2 M #3,4 F	2-7, 3-8 2-6, 2-4	Indoor: high density	#2M
G	M.mu.	#1 M #2,3 F	2-8 2-2, 3-0	Indoor: high density	#3F
H	M.mu.	#1,2 M #3 F	2-8, 2-1 2-5	Indoor: high density	#1M
I	M.mu.	#1 M #2,3,4 F	3-5 3-8, 3-0, 2-2	Indoor: high density	#2F → #3F[c]
D	M.fa.	#1 M #2,3 F	3-5 3-5, 4-3	Indoor: high density	#1M
E	M.fa.	#1 F #2,3 M #4· F	3-7 3-7, 3-4 3-7	Indoor: high density	#2M

[a] Species are *M. mulatta* or rhesus macaque and *M. fascicularis* or crab-eating macaque. [b] Age is expressed as years and months at start of THC administration. [c] The THC-treated subject was changed from #2F to #3F during the course of the study due to illness of #2F.

haviors for each group member. Assessment of behavioral changes over time for individuals was made by comparing mean frequencies from blocks of sequential daily observation sessions spaced at intervals of 3 months.

For each social group, each week of behavioral observations was followed by a week during which two 24-hour urine samples were taken from each group member, one under basal nonstimulated conditions, and one after injection of 4 units/kg ACTH. These urine samples were assayed for 24-hour excretion levels of cortisol, epinephrine, norepinephrine, MHPG (3-methoxy, 4-hydroxyphenylethylene glycol) and creatinine. Blood samples were taken from all subjects once a month for determination of plasma testosterone and cortisol levels. Blood samples were also taken as required from selected females for progesterone determination.[4]

RESULTS

Time course of drug effects on behavior

1) Immediate acute and short-term chronic effects: During the early stages of drug exposure, behavioral

---

[4] Steroids were assayed by standard competitive protein binding techniques; free catecholamines were determined by a semiautomated adaptation of the trihydroxyindole procedure; and MHPG was measured by gas chromatography.

effects were evident at 1 hour after drugging, maximal at 2 to 3 hours, and still apparent at 5 hours. During these postdrugging hours, all subjects showed both sedation and activation effects. Sedation was evidenced by increased frequencies and durations of sleep episodes and sitting and lying down, concomitant with suppression or elimination of exploratory behavior and high arousal responses such as cage shaking. At the same time, drugged subjects showed activation effects in increased active stereotypy such as pacing or flipping as well as in increased visual monitoring of their cagemates.

Aggressive behavior of dominant group members was generally decreased or eliminated after drugging, especially during the food competition test situations. An exception to this was noted for certain #1 ranking cagemates, who showed an increase in "irritable aggression" toward single specific nonfavorite cagemates. Drugged subordinates, however, tended to receive more aggression from nondrugged cagemates, resulting in some slight increase in the frequencies of their submissive responses. In general, recognition by cagemates of the altered social responsiveness of drugged subjects appeared to be one of the most sensitive indexes of drug effects. Thus, not only were drugged subordinates threatened and attacked more frequently, but drug-induced changes in dominant group members resulted in increased spontaneous aggression and competition for preferred food among their undrugged subordinate cagemates.

Drug-induced changes in affiliative behaviors appeared to reflect sedation effects of the drug: i.e., active affiliation in the forms of grooming and play was almost entirely eliminated, while passive affiliation such as huddling with preferred cagemates was enhanced for all subjects.

Self-directed behaviors involving tactile stimulation—such as self-groom, self-mouth and scratch—were also absent or greatly reduced during the postdrug hours.

In general, during early stages of drug exposure, sedation effects predominated. This was most evident when all members of a social group were drugged simultaneously. Under these conditions, active social interaction essentially ceased while sleep and passive affiliative behavior predominated.

2) Development of behavioral tolerance: During the period of 2 weeks to 2 months after the initiation of daily drug treatment, behavioral tolerance developed gradually, so that both activation and sedation effects were greatly reduced or eliminated. Although there was individual variability among drugged subjects, as tolerance developed all became essentially indistinguishable from undrugged cagemates during the immediate post-treatment hours; i.e., there was a disappearance of characteristic drug effects such as ptosis, hunched posture, or stereotypy, particularly under conditions of group activity. Under conditions of low level environmental stimulation or minimal social interactions, mild sedation effects were still discernible during the first few hours after drugging.

3) Emergence of irritable aggressiveness: Concurrent with the development of tolerance, an increased irritable responsiveness developed in drug-treated group members which took the form of slightly increased aggression for most of the THC-treated subjects. The social consequences of this effect were first manifest in shifts in the dominance struc-

TABLE 2. Dominance shifts versus length of chronic THC treatment for females in two mixed peer groups

Cage group	Dom. rank	Group members in order of rank[a]				
		Months of daily THC treatment[b]				
		0–1	2	6	7	8
OE	#1	1M*	1M*	1M*	1M*	1M*
	#2	2M	2M	2M	2M	5F*
	#3	3M	3M	3M	5F*	4F
	#4	4F	4F	4F	3M	6F
	#5	5F*	6F	5F	4F	[2M]c
	#6	6F	5F*	6F	6F	[3M]

		0–4	5	8	9	
OW	#1	1M	1M	1M	1M	
	#2	2M*	2M*	4F*———— 4F*		
	#3	3M	4F*	2M*	5F	
	#4	4F*	3M	3M	6F	
	#5	5F	5F	5F	[2M*]	
	#6	6F	6F	6F	[3M]	

[a] Group members are designated by predrug dominance rank and sex; M = male, F = female, * = THC-treated.   [b] THC was administered daily on preferred food at 2.4 mg/kg.   c [ ], removed from group to reduce aggression.

ture in the two six-membered peer groups, where the THC-treated female in each group rose stepwise in the dominance hierarchy to displace all cagemates but the alpha male, as shown in Table 2. These hierarchal shifts occurred at a time when the group dominance structure had been already weakened by intermale tensions following the first breeding season of the late-pubertal males. In both cage groups, these tensions ultimately resulted in recurring total group aggressive episodes which finally necessitated the removal of all males but the alpha male from each group, as shown in the last columns of Table 2. In these episodes of group aggression against the subordinate males, the THC-treated females were observed to play initiating roles, although the two THC-treated males did not react similarly.

The time course of chronic THC effects on social behavior is well illustrated by the data for Cage OE in Fig. 1, where it can be seen that the #5 female first dropped to rank #6 during the first months of THC treatment when sedation effects predominated. But as tolerance developed, she subsequently rose in rank in three stepwise dominance shifts over a period of several months.

In the indoor cage groups, high density stress and marked individual differences in age and size (see Table 1) resulted in more rigid dominance structures which precluded dominance shifts during the course of THC treatment. However, final summation of absolute frequencies of aggressive behaviors showed persistent increases in some form of aggression for five of six THC-treated group members.

Although the forms of increased aggressiveness varied among THC-treated group members, there was a consistent trend toward increases over predrug baseline frequencies in either spontaneous social interaction or competitive interaction or both, manifest primarily as hit, bite or chase contact aggression and occasionally as intense attack behaviors. Thus, the two THC-treated males in Cages D and F showed increases in hit-bite-chase behaviors in both spontaneous and competitive situations, while the treated males in Cages E and H increased aggression only in the food competition situation. Nondrugged cagemates in these same cage groups showed decreased or unchanged aggression frequencies. The single THC-treated subject which also showed no increase in aggression was the most subordinate female in Cage G, for whom aggressive behavior against larger male cagemates was precluded.

In the latter months of this study, seven females, including two THC-treated females, gave birth to offspring of which four survived. The one THC-treated mother with a surviving male offspring showed unique irritable rejection behaviors toward her infant which were not demonstrated by the two nondrugged mothers of surviving male infants, although all three mothers provided

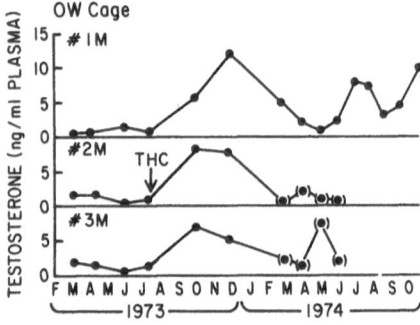

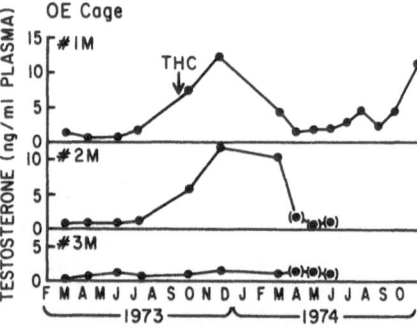

Figure 1. Seasonal plasma testosterone changes in males of outdoor caged peer groups. In each cage group only one of the three males was treated with THC as shown by the arrows.

access to adequate lactation to support a normal growth rate. In particular, analysis of behavior sequences recorded during the first 3 months after birth, showed the THC-treated mother tended to give more delayed responses or refusal to respond to infant protest vocalization, at the same time as the infant of the THC-treated mother gave more vocal protests to maternal acts.

Neuroendocrine function in THC-tolerant subjects

1) Reproductive hormones: plasma testosterone levels of post-pubertal males in the social groups studied appeared to be influenced by maturational, seasonal, and social factors but not by THC administration per se. Thus, in the indoor cage groups, monthly plasma testosterone determinations during the course of daily THC administrations showed the same degree of intra-individual fluctuation and similar ranges of inter-individual differences between drugged and nondrugged males. On a month-to-month basis, testosterone levels tended to be depressed in subordinate males subject to the social stress of more dominant male cagemates and elevated by the proximity of estrous females.

In both outdoor and indoor cage groups, THC-treated and nondrugged males showed comparable testosterone responses to the presence of estrous females. In the two outdoor peer groups, testosterone levels rose uniformly in five of the six pubertal males prior to their first fall breeding season, as shown in Fig. 1. Here the two THC-treated males are indistinguishable from three nondrugged peer cagemates in the timing and extent of their testosterone response. Similarly, THC-treated males in indoor cages showed threefold elevations in plasma testosterone levels when caged with or adjacent to an estrous female, with a return to lower levels following removal from proximity to this stimulus.

Long-term THC-treated females showed no consistent irregularity of menstrual cycles as indicated by vaginal bleeding. Similarly, premenstrual progesterone increases were found to follow a normal pattern when determined on semiweekly plasma samples over two menstrual cycles in two THC-tolerant females.

2) Pregnancies and offspring: Normal pregnancies occurred in the two THC-treated females in the low-stress outdoor cage groups. Births from these females were 1 to 2 months later than births from their four undrugged cagemates. Whether this was due to delayed conception or prolonged gestation could not be determined in the group breeding situation.

Offspring of the two THC-treated females were small but grossly normal at birth. One female infant died at birth with neurohistopathology of hydrocephalus and myocardial degeneration. The surviving male infant showed normal postweaning social interaction with peers, but persistent general hyperactivity.

3) Stress response hormones: During the course of long-term THC administration, 24-hour excretion levels of cortisol, epinephrine, norepinephrine and MHPG were determined for all 22 members of the indoor cage groups on a continuing basis so that triplicate determinations were made at 3 month intervals on a schedule paralleling the ongoing behavioral assessments. Although within each group there were marked individual differences in excretion levels of cortisol and catecholamines which correlated with social rank and/or social stress, there was no evidence for THC-induced alterations in these stress-response measures, either during the early stages of drugging or in the long-term tolerant drugtreated subject. Similarly, adrenal responsiveness to ACTH in each subject failed to show changes or interindividual differences attributable to THC-induced moderation of hypothalamic-pituitary responsiveness to overall environmental stimulation.

DISCUSSION AND CONCLUSIONS

The present study with THC in simian primates provides no evidence for endocrine derangement resulting from long-term chronic drug exposure. Specifically, at a drug level comparable to heavy daily cannabis use in man, there was no evidence for testosterone depression in males, for derangement of menstrual cycles, conception or lactation in females, or for changes in stress response systems.

The results of the current study do confirm, however, the uniqueness and subtlety of cannabis effects on social behavior. The initial short-term responses of naive subjects to the drug mimic many aspects of the "marihuana high," including sedation, activation and decreased social responsiveness. Surprisingly, long-term exposure to a constant daily level of THC did not produce effects in monkeys comparable to the apathy and "flatness of affect" of the marihuana syndrome; but it did reproduce the underlying irritability described by Kolansky and Moore, typically demonstrated in man as defensive aggressiveness or outbursts or irrational anger and verbal abuse (4).

In our test system, this change in affect was relatively subtle, and was demonstrated only in those social contexts which permitted its effective expression, or through analysis of extensive behavioral records to demonstrate quantitative increases in specific aggressive behaviors of individual group members. In the latter case, it was the inter-individual consistency of behavioral change among drugged subjects relative to nondrugged cagemates which argued most strongly for the validity and significance of the effect.

The phenomenon of behavioral tolerance is also a basic aspect of both THC and marihuana action, and is viewed as the result of neurologic adaptation to a chronically sustained drug level to reestablish basal neuronal function. Thus, an increase in drug level can reinitiate characteristic drug effects at the same time as it contributes to an increased level of tolerance. In our own studies, initiating or increasing daily THC levels produced sedation, activation, and depressed affect, while maintaining a constant daily drug level resulted in the development of tolerance to these effects plus increased irritability.

There is a similarity between these events and the phenomenon of dependency and withdrawal. In fact, irritability and aggressiveness have been reported during withdrawal from much higher drug levels in induced THC-dependency in primates (3). However, in our work with primates tolerant to lower drug levels we have seen no indication of dependency, as evidenced by a preference for drugged food, or of withdrawal, as evidenced by increased aggressiveness after cessation of daily THC administration.

Various contradictory aspects of marihuana effects in man may be manifestations of the interaction between varying levels of drug tolerance and the fluctuating drug levels resulting from episodic drug use. If immediate drug levels exceed the level of tolerance, sedation effects predominate; if drug levels drop below tolerance levels, irritability effects predominate. In cultures where cannabis exposure is characterized by more continual use of higher potency drug, the tolerance level would be expected to approximate dependence in the sense that high drug levels must continue to be maintained for normal neuronal functioning.

In considering the brain as a target organ for cannabis or THC, it is of interest to note that THC has been shown to be taken up preferentially in those limbic areas of the monkey brain that have been implicated in control of aggressive behavior, as well as in cortical areas that control functional intelligence

(7). Our limited observations of offspring of THC-treated females suggest that the developing nervous system may also be a sensitive target for low levels of THC which produce no apparent deleterious effects in the maternal host. The observed effects in two offspring which are suggestive of cannabis action—i.e., histopathologic hydrocephalus and behavioral hyperactivity—are of such a nature as to be difficult to substantiate as drug effects in human subjects. Continued observations in primate test systems can provide relevant information in this relatively unexplored area.

REFERENCES

1. BOWMAN, M., AND R. O. PIHL. Cannabis: Psychological effects of chronic heavy use. *Psychopharmacologia* 29: 159, 1973.
2. CAMPBELL, A. M. G., J. L. G. THOMPSON, M. EVANS AND M. J. WILLIAMS. Cerebral atrophy in young cannabis smokers. *Lancet* 1: 202, 1972.
3. DENEAU, G. A., AND S. KAYMAKCALAN. Physiological and psychological dependence to synthetic Δ^9-tetrahydrocannabinol (THC) in rhesus monkeys. *Pharmacologist* 13: 2, 1971.
4. KOLANSKY, H., AND W. T. MOORE. Toxic effects of chronic marihuana use. *J. Am. Med. Assoc.* 222: 35, 1972.
5. KOLODNY, R. C., W. H. MASTERS, R. M. KOLODNER AND J. TORO. Depression of plasma testosterone levels after chronic intensive marihuana use. *N. Engl. J. Med.* 290: 872, 1974.
6. MAUGH, T. H. Marihuana (II): Does it damage the brain? *Science* 185: 775, 1974.
7. MCISAAC, W. M., G. E. FRITCHIE, J. E. IDANPAAN-HEIKKILA, B. T. HO AND L. F. ENGLERT. Distribution of marihuana in monkey brain and concomitant behavioral effects. *Nature* 230: 593, 1971.

Effect of ambient levels of ozone on monkeys[1]

D. L. DUNGWORTH, W. L. CASTLEMAN,
C. K. CHOW, P. W. MELLICK, M. G. MUSTAFA,
B. TARKINGTON, AND W. S. TYLER

*California Primate Research Center
and Department of Veterinary Pathology
University of California, Davis, California 95616*

There is a considerable body of information concerning the experimental toxicity of ozone (16, 24), most of it relating to edemagenic concentrations above 1 ppm (2 mg/m^3). Few descriptions have been provided of pulmonary lesions produced by concentrations of 0.8 ppm and below (3–6, 14, 15, 23), which represents the range of oxidant activity encountered in photochemical smog such as occurs in the California South Coast Basin (2). Our combined functional, biochemical and morphological studies of ozone toxicity are designed to provide information about the pathogenesis and nature of the effects of ozone at levels attained in smoggy environments. Whereas the majority of experiments to date have been on rodents, our investigations entail the use of both monkeys and rats for direct interspecies comparison and more confident extrapolation of the results to man. This report focuses on the status of initial studies in monkeys and their implications.

EXPOSURE REGIMENS

Rhesus (*Macaca mulatta*) and bonnet monkeys (*M. radiata*) were selected for experiments after thorough screening of health records and confirmation of freedom from disease on the basis of clinical examination, radiography, hemograms and chemical analysis of serum.

During the seven daily exposure periods and for two prior conditioning periods, each monkey was housed individually in a 21 ft^3 chamber provided with chemical-biological-radiologically filtered air and maintained at 77 ± 3 F with relative humidity kept below 60%. Details of ozone generation and monitoring have been published (5). Control monkeys were treated the same way. Clinical, radiological and hematological parameters were monitored during the course of the experiment.

[1] This research was supported by Public Health Service Grants RR00169 and ESHL-00628.

Rhesus monkeys

Twenty-four female rhesus monkeys, 3.2 to 5.3 years of age and weighing 3.8 to 6.3 kg, were used for the protocol outlined in Table 1. All had been born and raised in the colony and on the basis of past experience were expected to be free from lung mites.

Significant clinical changes were limited to higher respiratory rates ($P = 0.01$) for monkeys in both exposure groups during the first three periods. Data recorded during the seventh hour of each 8-hour exposure period were analyzed by a Student-Newman-Keul's test following analysis of variance. Respiratory rates were most increased in the first exposure period and were progressively less affected thereafter so that by the fourth period there was no significant effect.

Bonnet monkeys

Adult female bonnet monkeys, approximately 6 to 15 years old and weighing 3.4 to 7.0 kg, were used in the protocol outlined in Table 1. Significant clinical changes were limited to higher respiratory rates ($P = 0.01$) for the first three exposures in the two monkeys exposed to 0.5 ppm, although they were not so pronounced as for the rhesus monkeys exposed to 0.5 ppm.

Terminal procedures

Immediately following the seventh exposure, the monkeys were killed by intravenous administration of pentobarbital sodium and exsanguination, and the lungs were removed. For rhesus monkeys, the right lung was fixed by airway perfusion at constant pressure of 30 cm of water with Karnovsky's fixative diluted 1:4.5 (550 m0sm) with 0.2 M cacodylate buffer. Parenchyma of the left lung was dissected away from major airways and homogenized for biochemical determinations. Subsequently, for bonnet monkeys, it was found to be more convenient to fix the left lung for morphology and use the right lung for biochemical determinations.

BIOCHEMICAL ALTERATIONS

Lung homogenates and various subcellular fractions, i. e., mitochondria, microsomes and cytosol, were prepared according to procedures described previously (19, 20). Biochemical parameters that had been found to be indexes of damage or response in lungs of rats exposed to the same concentrations of ozone (8–12, 19, 20) were determined. Typical results for two of the marker enzyme activities, succinate oxidase and glutathione peroxidase, are given in Table 2. In exposed monkeys, there was increased activity of both enzymes, with a greater effect from higher concentrations of ozone. Linear regression analysis showed a significant correlation ($P < 0.05$) between ozone concentration and augmentation of activity of both enzymes for rhesus and bonnet monkeys. The same was true for other enzymes measured: glucose-6-phosphate dehydrogenase, NADPH-cytochrome c reductase, GSSG reductase,

TABLE 1. Intermittent ozone exposure for 8 hours a day on 7 consecutive days

	Number of monkeys	Ozone concn., ppm
Rhesus	6	0.8 ± 0.05
	6	0.5 ± 0.04
	12	0
Bonnet	2	0.5 ± 0.03
	3	0.35 ± 0.02
	4	0.2 ± 0.02
	4	0

TABLE 2. Increased activities of succinate oxidase and glutathione peroxidase in lungs of monkeys following exposure to ozone for 8 hours a day on 7 consecutive days

Ozone concn., ppm	Succinate oxidase[a]		Glutathione peroxidase[b]	
	Rhesus	Bonnet	Rhesus	Bonnet
0	3.0 ± 0.2^c	3.6 ± 0.3	20 ± 3	38 ± 4
0.2	—	4.0 ± 0.3	—	40 ± 5
0.35	—	4.4 ± 0.4	—	43 ± 5
0.5	3.3 ± 0.3	5.4 ± 0.4	22 ± 4	46 ± 3
0.8	3.7 ± 0.3	—	23 ± 3	—

[a] n moles O_2/min per mg protein.   [b] n moles NADPH oxidized/min per mg protein.   [c] mean and standard deviation.

acid phosphatase, and β-N-acetyl-glucosaminidase.

MORPHOLOGICAL FINDINGS

There were neither gross lesions nor significant differences in lung to body weight ratios and displacement volumes of lungs from exposed animals.

Trachea, bronchi, terminal and respiratory bronchioles and alveolar parenchyma were specifically selected from cranial, middle, and caudal lobes and examined by correlated techniques of light microscopy and scanning and transmission electron microscopy. Airways were longitudinally bisected; one side was prepared for light microscopy and scanning electron microscopy, and the other side was embedded in molds for 1–2 micron sectioning of large faces and subsequent transmission electron microscopy as described previously (17, 21).

The most obvious and consistent pulmonary lesion in all exposed monkeys occurred in respiratory bronchioles. The extent and severity of damage, but not its nature, varied with the level of ozone exposure. There was a progressive decrease in amount of damage from proximal to distal orders of bronchioles and the farthest point reached depended on dosage. With 0.2 ppm, the lesion was largely limited to the proximal generation of respiratory bronchioles whereas with 0.8 ppm the damage extended to involve, minimally, proximal portions of alveolar ducts. The lesion was most readily demonstrated by scanning electron microscopy, especially at the lower dose levels. Luminal surfaces and alveolar openings of respiratory bronchioles were coated by numerous macrophages, occasional neutrophils and eosinophils, and small quantities of debris (Figs. 1 to 4). There were both hyperplasia and hypertrophy of bronchiolar epithelium which resulted in replacement of the usual nonciliated, cuboidal and squamous cells by low columnar cells rich in organelles (Fig. 5). Degenerative cytoplasmic changes were commonly observed at the two highest dose levels; necrosis and sloughing were seen infrequently.

Epithelium lining alveoli in the walls of respiratory bronchioles was thickened by increased numbers of alveolar type 2 cells. Some of the cells, however, were flattened and sparsely populated by microvilli and lamellar inclusions. They resembled cells intermediate between types 1 and 2 epithelial cells similar to those described in NO_2 (13) and oxygen-induced lesions (1). Connective tissue in the walls of respiratory bronchioles was often slightly edematous, except at 0.2 ppm, and there was a light infiltration of mixed inflammatory cells.

The trachea and bronchi of monkeys also were affected at all levels of exposure. Again, lesions were most easily detected by scanning electron microscopy as randomly-scattered foci predominantly involving ciliated cells. Cilia were either absent or short and blunt. Transmission electron microscopy revealed degenerative changes and evidence of ciliogenesis. Mucus-producing cells were less affected, but were seen to be flattened and covered with microvilli and coarse nodular surface projections.

DISCUSSION

Several important inferences can be drawn from these early studies and comparable ones in rats. A 7-day intermittent exposure for 8 hours a day to as low as 0.2 ppm of ozone consistently produces mild damage to the conducting airways and respiratory bronchioles of bonnet monkeys. This level of exposure to oxidant can be experienced fairly frequently between 9:00 AM and 5:00 PM during summer months in parts of the South Coast Basin (2). We have found the threshold of detectable biochemical and morphological effect of the same intermittent regimen in rats to be 0.1 ppm (manuscript in preparation). Since the threshold level of effect in bonnet monkeys is less than 0.2 ppm, and is considerably closer to that of the rat than was perhaps to be expected, it follows that it is much more likely that similar lesions can be produced in previously unexposed persons at concentrations down close to the 0.2 ppm level.

With regard to the choice of an 8-hour intermittent exposure, we

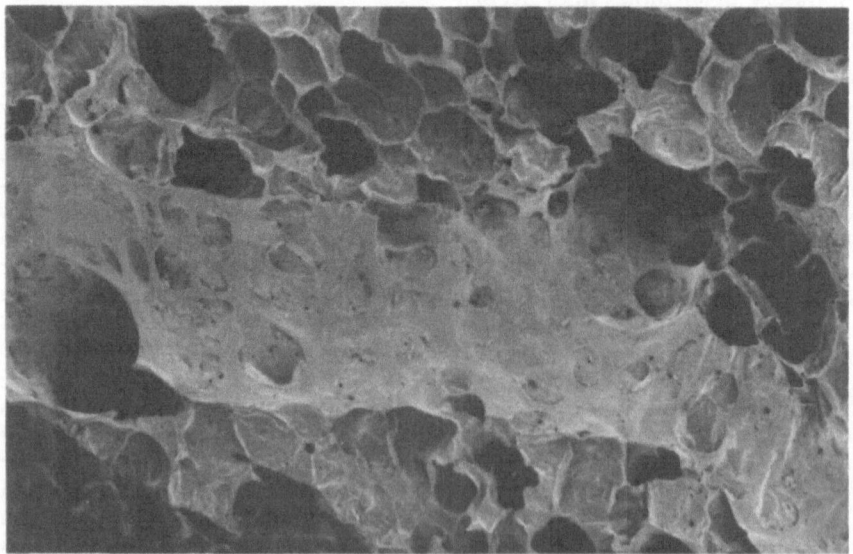

Figure 1. Scanning electron microscopy of respiratory bronchiole from control bonnet monkey. The bronchiolar wall has numerous alveolar outpocketings which contain only a few alveolar macrophages. ×52.

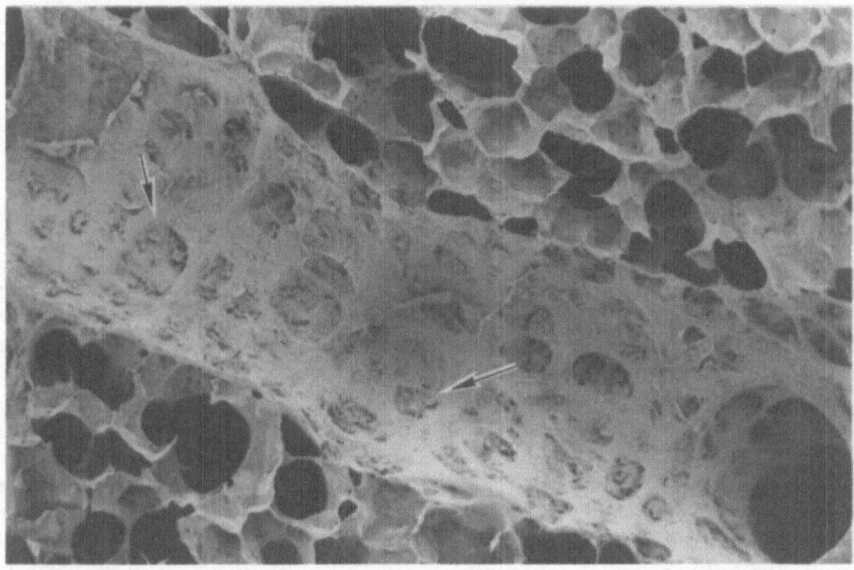

Figure 2. Proximal area of respiratory bronchiole from bonnet monkey exposed to 0.20 ppm ozone 8 hours a day for 7 days. The luminal surface is covered by a thin layer of exudate. Alveoli within the bronchiolar wall (arrows) contain aggregations of cells which are predominantly alveolar macrophages. Interalveolar septa surrounding the airway are not involved in the inflammatory process. ×62.

have found in rats on quantitative biochemical grounds (8) and by morphological examination (manuscript in preparation) that the severity of damage produced by an 8-hour exposure on each of 7 days was similar to that produced by continuous 7-day exposure. This is consistent with the finding that an 8-hour exposure to 0.9 ppm ozone elicits a pulmonary lesion which when examined 40 hours after the end of exposure is morphologically indistinguishable from the lesion produced by 48 hours of continuous exposure to the same concentration (23). The importance of this is that at the low concentrations encountered in smog the amount of damage is much more dependent on the highest concentration achieved than it is on the mean exposure over a 24-hour period. Whether the daily exposure period can be reduced below 8 hours without substantially

lessening the insult remains to be determined. The concept of concentration × time being constant in terms of amount of damage produced (25)

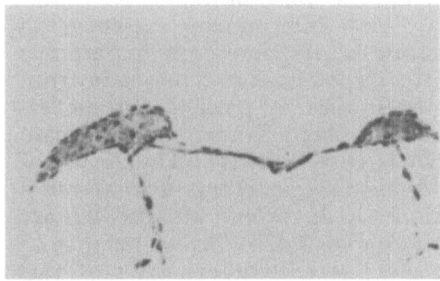

Figure 3. Respiratory bronchiole from control rhesus monkey. Shallow alveolar outpocketings are interspersed between broad bands of smooth muscle which are covered by a single layer of low cuboidal nonciliated epithelial cells. (Epoxy-embedded 1 μm section; methylene blue, azure II, and basic fuschin, ×190).

is clearly invalid for the low-dose effects.

As has been observed in other species (4, 5, 15, 22, 23), the most prominent lesion in ozone-exposed monkeys is centriacinar, that is at the junction of conducting airways and exchange tissue. The precise site of major involvement is dependent on anatomic features of the species exposed and concentration of ozone. In both monkeys and rats, the higher the concentration of ozone the farther the lesion extended into the pulmonary parenchyma. At the levels studied, the principal site of damage in monkeys was the respiratory bronchiole whereas in rats it was the alveolar duct. Since the degree of development of respiratory bronchioles in man more nearly resembles that of monkeys than of rats (7), the monkey is the better indicator of where damage might occur in man. The consequences of focal lesions observed in major airways, unless of significant promotional effect for pulmonary carcinogenesis, are unlikely to be as serious as those of the lesions in small airways which can exacerbate or lead to chronic bronchitis and emphysema.

The determination of the subsequent evolution of lesions seen after a 7-day intermittent exposure to ozone is of paramount importance if extrapolation is to be made from our studies to possible chronic effects of photochemical smog on man. We have recently found by the use of biochemical markers such as mentioned in this report, with preliminary morphological confirmation, that in daily intermittent exposure of rats through 30 days the damage reaches a peak by the third or fourth day and levels off thereafter. The first mention of such a plateau was by Mittler (18). Stevens et al. (23) also reported the time to full morphologic development of the lesion under continuous exposure to be 3 days. Changes evoked during the first 3 to 4 days bring about a state of adaptation whereby continued exposure results in little or no more immediate damage and the time frame of subsequent changes becomes much larger. Present evidence suggests the process of adaptation is due to shifts in cell population and modifications in biochemical constituents (6). The former at the alveolar level is achieved by replacement of highly vulnerable type 1 epithelial cells by resistant type 2 cells (5, 23), the latter at least in part by increases in activities of enzymes such as glutathione peroxidase that can bolster antioxidant defense mechanisms (9, 10). It is important to note that the phenomenon of tolerance to ozone as it is used to describe protection against lethal edemagenic doses of ozone (24, 25) should not be confused with adaptation in the sense of the lung reaching an early plateau of damage in response to low levels of ozone. Although overlapping sets of pathogenetic factors are involved in the

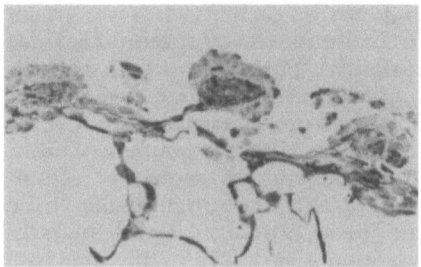

Figure 4. Respiratory bronchiole from rhesus monkey exposed to 0.5 ppm O$_3$, 8 hours a day for 7 days. Alveolar outpocketings contain numerous macrophages, debris and occasional neutrophils. Bronchiolar epithelial cells are much larger and more numerous than normal. (Epoxy-embedded 1 μm section; methylene blue, azure II, and basic fuschin, ×190).

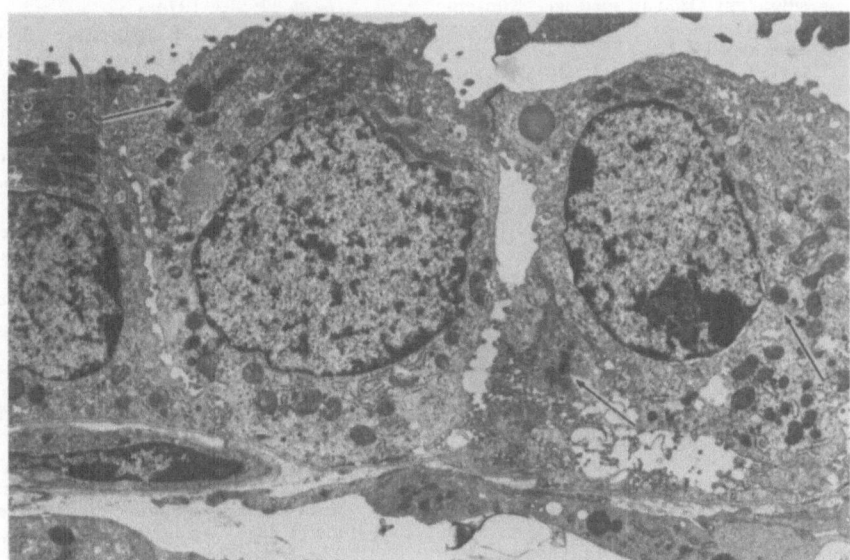

Figure 5. Transmission electron microscopy of bronchiolar epithelium from respiratory bronchiole of rhesus monkey exposed to 0.5 ppm O_3, 8 hours a day for 7 days. The cell on the right has numerous degenerative changes including cytoplasmic vacuolation, separation from the adjacent cell except at the tight junction, and several dense membrane-bound cytoplasmic granules resembling lysosomes (arrows). Amorphous exudate in the airway lumen is present above this cell. The cell in the center also contains dense cytoplasmic granules (arrow) and increased numbers of mitochondria and profiles of granular endoplasmic reticulum. Two granulocytes are present immediately beneath the basal lamina. (Uranyl acetate and lead citrate, ×6,660).

two phenomena and neither is specific for ozone, the overall implications are quite different.

The chronic effects of exposure to the ambient levels of ozone as featured in this report depend on the balance between the efficacy of adaptation and degree of insult as it affects the probability of continuation of low-grade inflammatory stigmata. The latter conceivably can lead to chronic bronchitis and emphysema or have a cocarcinogenic effect.

Our major current goals are to establish how the balance of adaptation and insult affects evolution of pulmonary damage in the range of ozone concentration from 0.2 to 0.8 ppm, and to seek the mechanisms involved. Studies in monkeys are par-ticularly pertinent relative to man because, like man, they have a well-developed system of respiratory bronchioles.

REFERENCES

1. ADAMSON, I. Y. R., AND D. H. BOWDEN. The Type 2 cell as progenitor of alveolar epithelial regeneration. A cytodynamic study in mice after exposure to oxygen. *Lab. Invest.* 30: 35, 1974.
2. *Basic Tabulation, California Air Monitoring Stations,* California Air Resources Board, Sacramento, California, 1971.
3. BOATMAN, E. S., S. SATO AND R. FRANK. Acute effects of ozone on cat lungs. II. Structural. *Am. Rev. Respirat. Diseases* 110: 157, 1974.
4. BRUCH, J., AND H-W. SCHLIPKÖTER. Veränderungen der Lungenalveolen bei der Maus nach chronischer Expo-

sition mit Ozon in niedriger Konzentration. *Virchows Arch. Pathol. Anat.* 358: 355, 1973.

5. CASTLEMAN, W., L., D. L. DUNGWORTH AND W. S. TYLER. Cytochemically detected alterations of lung acid phosphates reactivity following ozone exposure. *Lab. Invest.* 29: 310, 1973.

6. CASTLEMAN, W. L., D. L. DUNGWORTH AND W. S. TYLER. Histochemically detected enzymatic alterations in rat lung exposed to ozone. *Exptl. Molec. Pathol.* 19: 402, 1973.

7. CASTLEMAN, W. L., D. L. DUNGWORTH AND W. S. TYLER. Intrapulmonary airway morphology in three species of monkeys. A correlated scanning and transmission electron microscopic study. *Am. J. Anat.* 142: 107, 1975.

8. CHOW, C. K., C. J. DILLARD AND A. L. TAPPEL. Glutathione peroxidase system and lysozyme in rats exposed to ozone or nitrogen dioxide. *Environ. Res.* 7: 311, 1974.

9. CHOW, C. K., AND A. L. TAPPEL. Activities of pentose shunt and glycolytic enzymatic protective mechanism against lipid peroxidation damage to lungs of ozone-exposed rats. *Lipids* 7: 518, 1972.

10. CHOW, C. K., AND A. L. TAPPEL. An enzymatic protective mechanism against lipid peroxidation damage to lungs of ozone-exposed rats. *Lipids* 7: 518, 1972.

11. DELUCIA, A. J., P. M. HOQUE, M. G. MUSTAFA AND C. E. CROSS. Ozone interaction with rodent lung: Effect on sulfhydryls and sulfhydryl-containing enzyme activities. *J. Lab. Clin. Med.* 80: 559, 1972.

12. DILLARD, C. J., N. URRIBARRI, K. REDDY, B. FLETCHER, S. TAYLOR, B. DE LUMEN, S. LANGBERG AND A. L. TAPPEL. Increased lysosomal enzymes in lungs of ozone-exposed rats. *Arch. Environ. Health* 25: 426, 1972.

13. EVANS, M. J., L. J. CABRAL, R. J. STEPHENS AND G. FREEMAN. Renewal of alveolar epithelium in the rat following exposure to NO_2. *Am. J. Pathol.* 70: 175, 1973.

14. FREEMAN, G., L. T. JUHOS, N. J. FURIOSI, R. MUSSENDEN, R. J. STEPHENS AND M. J. EVANS. Pathology of pulmonary disease from exposure to interdependent ambient gases (nitrogen dioxide and ozone). *Arch. Environ.*

Health 29: 203, 1974.

15. FREEMAN, G., R. J. STEPHENS, D. L. COFFIN AND J. F. STARA. Changes in dogs' lungs after long-term exposure to ozone. Light and electron microscopy. *Arch. Environ. Health* 26: 209, 1973.

16. JAFFE, L. S. The biological effects of ozone on man and animals. *Am. Ind. Hyg. Assoc. J.* 28: 267, 1967.

17. LOWRIE, P. M., AND W. S. TYLER. Selection and preparation of specific tissue regions for TEM using large epoxy-embedded blocks. *31st Ann. Proc. Electron Microscopy Soc. Amer.*, edited by C. J. Arceneaux. New Orleans, La. 1973.

18. MITTLER, S., M. KING AND B. BURKHARDT. Toxicity of ozone. III. Chronic toxicity. *A.M.A. Arch. Ind. Health* 15: 191, 1957.

19. MUSTAFA, M. G., AND C. E. CROSS. Effects of short-term ozone exposure on lung mitochondrial oxidative and energy metabolism. *Arch. Biochem. Biophys.* 162: 585, 1974.

20. MUSTAFA, M. G., A. J. DELUCIA, G. K. YORK, C. ARTH AND C. E. CROSS. Ozone interaction with rodent lung. II. Effects on oxygen consumption of mitochondria. *J. Lab. Clin. Med.* 82: 357, 1973.

21. NOWELL, J. A., AND W. S. TYLER. Scanning electron microscopy of the surface morphology of mammalian lungs. *Am. Rev. Respirat. Diseases* 103: 313, 1971.

22. SCHEEL, L. D., O. J. DOBROGORSKI, J. T. MOUNTAIN, J. L. SVIRBELY AND H. E. STOKINGER. Physiologic, biochemical, immunologic and pathologic changes following ozone exposure. *J. Appl. Physiol.* 14: 67, 1959.

23. STEPHENS, R. J., M. F. SLOAN, J. J. EVANS AND G. FREEMAN. Early response of lung to low levels of ozone. *Am. J. Pathol.* 74: 31, 1974.

24. STOKINGER, H. E., AND D. L. COFFIN. Biologic effects of air pollutants. *Air Pollution*, 2nd ed., edited by A. C. Stern. New York: Academic, 1968, vol. 1, p. 445.

25. STOKINGER, H. E. Evaluation of the hazards of ozone and oxides of nitrogen. Factors modifying toxicity. *A.M.A. Arch. Ind. Health* 15: 181, 1957.

Response of the nonhuman primate to polychlorinated biphenyl exposure[1]

J. R. ALLEN

Regional Primate Research Center
University of Wisconsin Medical School, and Food Research Institute
University of Wisconsin, Madison, Wisconsin 53706

The polychlorinated biphenyls (PCBs) have been used extensively for industrial purposes during the past 40 years. These compounds are extremely stable, not hydrolyzed by water, acid or alkali, and are able to withstand temperatures up to 650 C without disintegrating. These properties make them ideal for use in adhesives, paints, varnishes, printing inks, and as general fillers. Since they do not conduct electricity, they have found widespread use in electrical equipment such as transformers. However, it was only during the past decade that the health significance of these compounds in man and lower animals has been brought to public attention. Prior to this time only workers exposed to the PCBs during the production process (15) or workers who were in direct contact with manufactured products containing PCBs (10) occasionally developed a skin rash on exposed portions of the body. It was not until 1966 that Jensen (14) reported the presence of PCBs in the tissues of wildlife suffering from what was thought to be DDT intoxication. As a result of these findings other investigators began to evaluate the magnitude of environmental contamination of this newly discovered pollutant.

The public health significance of the PCBs was brought to the forefront when over 1,000 Japanese consumed rice oil that had been contaminated with a PCB mixture. In this outbreak the exposed persons developed chloracne and subcutaneous edema. Infants born to exposed mothers were small, exhibited discolored skin and had an abnormal eye discharge. Many of the symptoms and lesions that developed in those persons exposed to the PCBs have persisted (17).

The properties that make the PCBs ideal for commercial usage also enhance their resistance to degradation in the environment. The magnitude of PCB contamination is exemplified

[1] Supported in part by Public Health grants RR-00167, ES-00472 and CA-13288 from the National Institutes of Health, and the University of Wisconsin Sea Grant Program. Primate Center Publication no. 14–026.

Abbreviations: PCB, polychlorinated biphenyl; TCB, 2,5,2′,5′-tetrachlorobiphenyl.

73

by their presence in coho salmon, milk fat, poultry, eggs and fish (16). Detectable levels of PCBs are also present in over 30% of randomly sampled inhabitants of the United States (22).

In the following presentation a detailed description of the biological responses of nonhuman primates to various levels of PCBs will be presented, thus establishing this animal species as an ideal model for evaluating PCB intoxication in man. Data are also presented which show that levels of PCBs presently permitted in some foods destined for human consumption will produce ill effects in nonhuman primates within a relatively short period.

Effects on nonhuman primates of high level exposure to PCBs

When adult *Macaca mulatta* monkeys were given diets containing 100 and 300 ppm of a polychlorinated biphenyl (Aroclor 1248, Monsanto, St. Louis, Mo.) for periods ranging from 2 to 3 months, they experienced extreme morbidity within 1 month and mortality approaching 100% within 3 months. Although the signs and lesions that developed in these animals became apparent equally as soon in those receiving the lower dosage of PCBs, the total intake varied from 0.8 to 1.0 g for the 100 ppm group to 3.6 to 5.4 g for the 300 ppm group (5). There was a gradual weight loss experienced by all of the PCB-fed animals, with the 300 ppm group having a 25% decrease in body weight during the 3-month period of exposure. Within 3 weeks to 1 month the animals had lost a majority of their hair from the face, head, and neck, and the mouth and eyelids were edematous (6). In addition, there was a loss of eyelashes, excessive lacrimation, and conjunctival congestion.

Small comedones were particularly obvious around the mouth and on the cheeks and neck (Fig. 1).

These animals also showed a gradual decrease in hemoglobin and hematocrit. Although the total white cell count was not altered appreciably, there was a decrease in the number of circulating lymphocytes and a concomitant increase in neutrophils. Decreases in the serum proteins were related to a reduction in the level of albumin. Reduced total serum lipids, cholesterol, and triglycerides accompanied the altered serum protein (1).

The major microscopic changes in the skin of the affected animals were the development of large intrafollicular keratin cysts and epithelial hyperplasia of the hair follicles par-

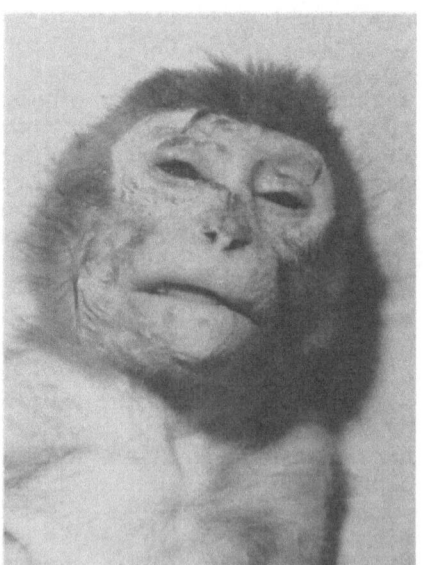

Figure 1. Facial edema is depicted in a monkey given 100 ppm Aroclor 1248 for 1 month. Swollen lips and eyelids and absence of eyelashes are common features of these animals. Small comedones are discernible on the upper lip, nose and forehead.

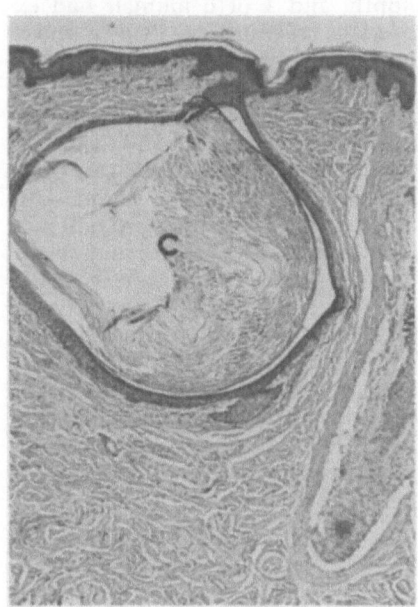

Figure 2. Large keratin cysts (C) in the hair follicles are common on the face of monkeys exposed to PCBs. Tissue from animal given 100 ppm Aroclor 1248 for 3 months. Hematoxylin and eosin stain; ×60.

ticularly of the face with the eyelids being most severely affected (Fig. 2). In addition, in many instances the tissue surrounding these affected hair follicles was edematous and contained acute inflammatory cells (1, 6). There were also major modifications in the morphological features of the stomach, particularly in the fundic and pyloric regions (Fig. 3). The markedly thickened gastric mucosa contained numerous large cysts filled with mucin and lined by epithelium. Penetration of the muscularis mucosa into the underlying submucosa by the glandular epithelium was a common feature of these affected stomachs. Cystic areas similar to those present in the mucosa were larger and more abundant in the submucosa. In addi-

tion, ulcerations of the gastric mucosa developed in areas of eroded epithelium and where large mucinous cysts had ruptured (6). This hyperplastic gastritis has persisted in monkeys for over 1 year following the discontinuation of PCB exposure (5).

Ingestion of the PCBs by monkeys caused a decided hypertrophy of the liver which in many instances amounted to over a twofold increase in size. Electron microscopically, the hepatocytes of these livers showed a decided increase in smooth endoplasmic reticulum (Fig. 4). In addition, there were segments of endoplasmic reticulum that formed elaborate concentric membrane arrays. Terminally, the livers of these animals continued to have enlarged hepatocytes. However, there was a rearrangement of the proliferated smooth endoplasmic reticulum in these cells. Instead of being distributed throughout the cytoplasm it was arranged in well circumscribed packets of closely associated membranes (Fig. 5) (1).

Liver biopsies taken from these animals during the course of the experiments showed a decided proliferation of the smooth endoplasmic reticulum of the hepatic cells. Biochemically, the liver homogenates contained a decreased level of DNA (mg/g liver) and RNA (mg/mg DNA) and increased activity of microsomal mixed function oxidases. Terminally, as the morphological features of the proliferated endoplasmic reticulum were modified there was a decided decrease in the activity of the mixed function oxidases (1). Such changes in the endoplasmic reticulum have been designated by Hutterer et al. (13) as hyperplastic, hypoactive endoplasmic reticulum.

At necropsy, tissues from these animals were analyzed by gas liquid chromatography for their PCB content. The major storage site for the PCBs was in the adipose tissue and to a lesser extent in the liver, adrenals, and pancreas. Regardless of the site of deposition the higher chlorine isomers predominated in the tissues (Fig. 6).

Low level PCB exposure

Adult female rhesus monkeys have been fed diets containing 2.5, 5.0 and 25.0 ppm of Aroclor 1248 over periods ranging from 2 months for the higher PCB level to 1 year for the two lower doses (4, 8). Animals on the 25 ppm dose developed facial edema, alopecia and acne within 1 month, and 1 of 6 animals had expired as a result of PCB intoxication 2 months after having been removed from the experiment diet. The total intake of PCBs during the experiment ranged from 250 to 400 mg per animal. As was the case with the higher doses, these animals developed anemia, hypoproteinemia, bone marrow hypoplasia and severe hypertrophic hyperplastic gastritis. PCB concentrations in samples of subcutaneous adipose tissue obtained from the animals averaged 127 $\mu g/g$ fat at the time the experimental diet was discontinued. Eight months later the PCB content of the adipose tissue had decreased to 34 $\mu g/g$ fat.

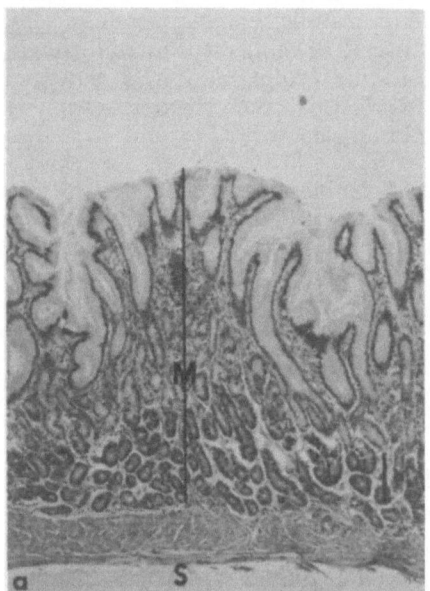 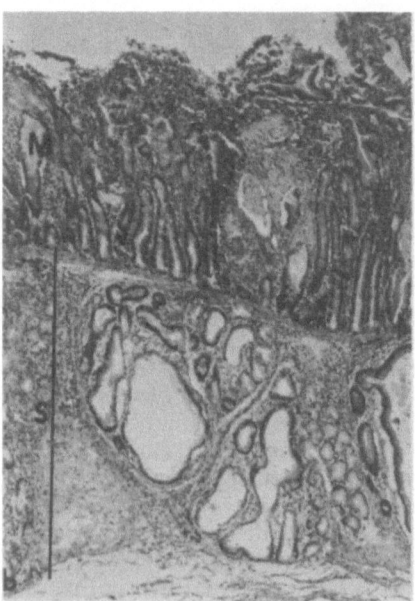

Figure 3. a) The morphological features of the normal gastric mucosa (M) of the monkey are depicted. The muscularis mucosa is intact and glandular elements in the submucosa (S) are absent. Hematoxylin and eosin stain; ×28. b) Hypertrophy and hyperplasia involving primarily the mucous secreting cells of the gastric mucosa (M) developed in monkeys fed diets containing Aroclor 1248 ranging from 25 to 300 ppm for 2 to 3 months. Penetration of the muscularis mucosa by the glandular epithelial elements and disruption of the underlying submucosa is a common feature of these affected stomachs.

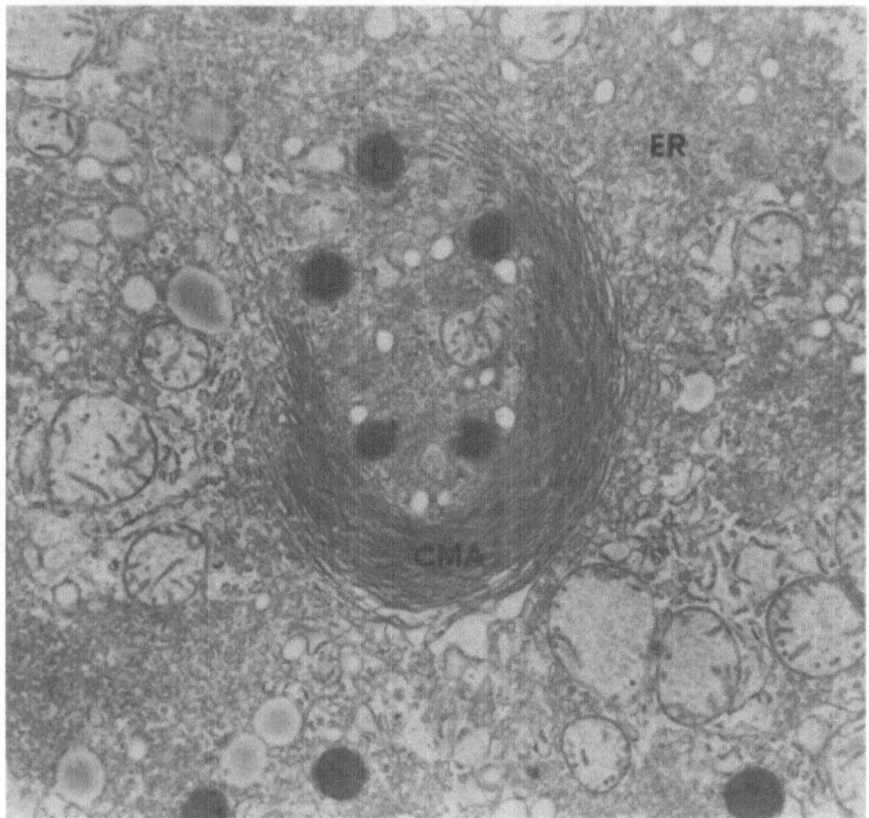

Figure 4. Hepatocyte of a monkey that had received 100 ppm Aroclor 1248 in the diet for 1 month. Abundant smooth endoplasmic reticulum (ER) and concentric membrane arrays (CMAs) develop in the cytoplasm of these cells. Numerous lysosomes (L) are dispersed throughout the cytoplasm. Uranyl acetate stain; ×14,280.

However, the surviving animals continued to show clinical signs and lesions of PCB intoxication 2 years following PCB exposure. Infants born to these PCB-fed females were small and contained detectable levels of PCB in their tissues (Table 1).

When adult rhesus monkeys were fed diets containing PCB equal to and one-half that permitted in some foods destined for human consumption (2.5 and 5.0 ppm), the females of the group developed periorbital edema, alopecia, erythema and acneform lesions that involved the face and neck within 1 to 2 months. Even though the males consumed more PCB, they exhibited only moderate periorbital edema and erythema. During the course of the experiment the females on the higher PCB dose consumed 505 mg while those on the lower dose ingested 270 mg. The males which received only the higher dose consumed 675 mg of PCB. After having established a relatively steady tissue level of PCB at 6 months, the females were bred to control males. Of the animals that conceived, 12.5% of the 5 ppm group and 62.5% of the 2.5 ppm group were able to deliver live infants as compared to 100% of the control group. However, the conception rate of females bred to PCB-fed males was equally as great as that in females bred to control males. Throughout the course of the experiment the females that consumed the PCBs had higher levels of urinary ketosteroids and a decided increase in the length of

menses and amount of menstrual bleeding (8).

Absorption, metabolism, tissue deposition and excretion of PCB

When adult rhesus monkeys were given a single oral dose of Aroclor 1248 and the urine and feces evaluated chromatographically for 14 days, over 90% of the compound was absorbed from the gastrointestinal tract. Ten percent of the original dose was detected in the excreta

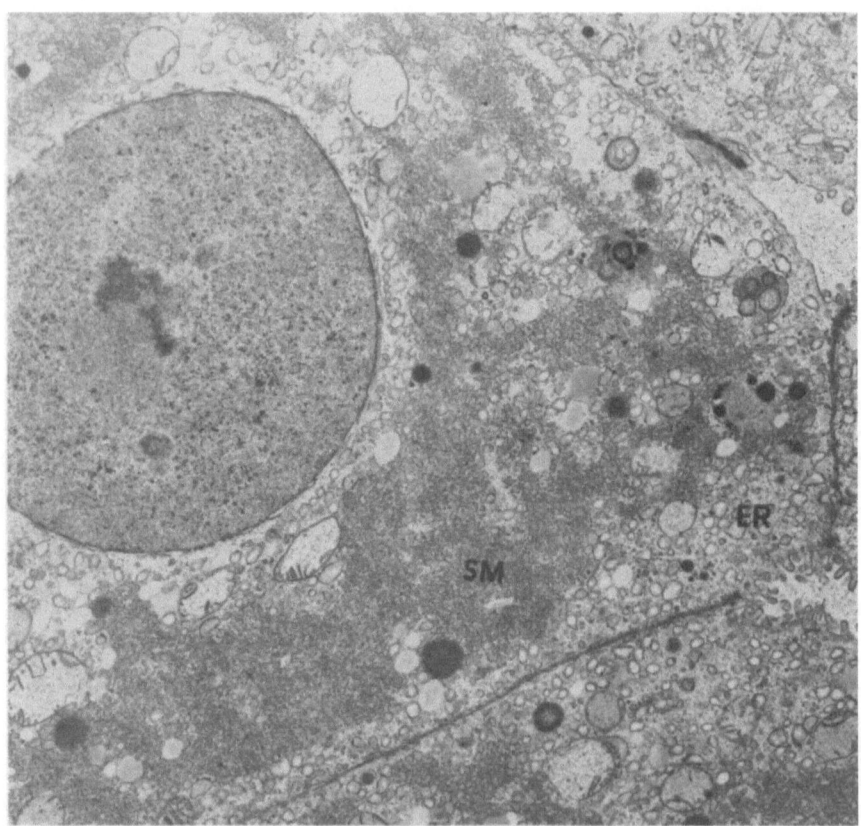

Figure 5. A very fine network of smooth surfaced membranes (SM) is present in the cytoplasm of hepatocytes from monkeys fed Aroclor 1248 in the diet for 3 months. Note the vesicular appearance of the remaining portions of the endoplasmic reticulum (ER). Contrast these membranes with those of Fig. 4. Uranyl acetate stain; ×5930.

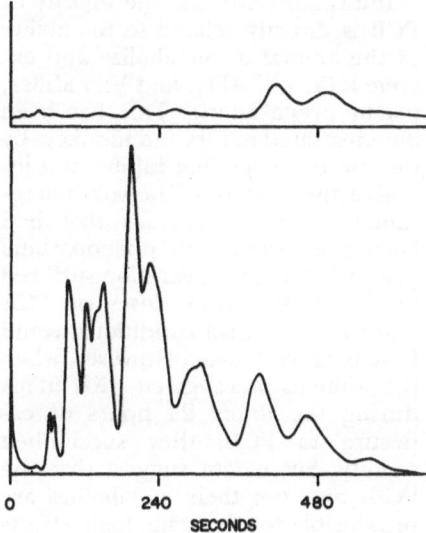

0   240   480
SECONDS

Figure 6. GLC-EC tracing of Aroclor 1248 (lower); the residue in the fat of a monkey given a diet containing 25 ppm Aroclor 1248 for 2 months (upper).

within 14 days with the greatest percentage occurring between the second and eighth day (7). When a tritiated PCB isomer (2,5,2',5'-tetrachlorobiphenyl) (TCB) was given to monkeys, less than 5% of the tritium had been eliminated in the excreta within 72 hours (12). Analysis of the excreta indicated that direct hydroxylation was not a major metabolic pathway for TCB in the monkey as is the case in the rat (20). Instead, hydroxylation occurred through the arene oxide intermediate (12). The TCB which remained in the body of these animals was primarily associated with the tissue protein and nucleic acids and serum albumin of the blood. In addition, the analysis of hexane extracts of liver and skin homogenates revealed that over 95% of the TCB present was in an unmetabolized form (12). When phenobarbital and SKF-525A were given to adult monkeys to increase and decrease

their hepatic microsomal enzyme activities respectively, the response of these animals to TCB was decidedly different. The animals pretreated with SKF 525A died within 48 hours on a dose of TCB that produced no obvious ill effects in animals that had been pretreated with phenobarbital (8).

COMMENTS

In experiments that have been conducted to determine the effects of polychlorinated biphenyls on nonhuman primates it has been established that the signs and lesions that appear in these animals are similar to those that occur in man exposed to similar levels of PCBs. Acne, subcutaneous edema, particularly about

TABLE 1. PCB content of tissues obtained from mother and infant monkey following consumption of Aroclor 1248[a] by mother prior to pregnancy[b]

Organ	Weight, gm	PCB content, μg/gm tissue
Mother		
Liver		56.3
Fat		50.0
Placenta		0.9
Infant		
Fat		27.70
Adrenals	0.2	24.40
Muscle		0.98
Stomach		0.55
Small intestine		0.52
Skin		0.31
Brain	52.9	0.29
Lung	6.1	0.21
Kidney	2.8	0.10
Large intestine		0.08
Liver	11.9	0.01

[a] Monsanto Co., Inc., St. Louis, Mo.   [b] Adult on diet containing 25 ppm PCB for 2 months and subsequently placed on a control diet for 2.5 months before breeding.

the face, and conjunctivitis along with excessive secretion of the meibomian glands were constantly present in PCB exposed man and lower primates (6, 17). The more severely affected animals have a decrease in erythrocytes, reduced hemoglobin and a leukocytosis. The most debilitating lesions in the monkeys were the severe gastric mucosal hyperplasia and ulceration. Whether similar changes occur in the stomach of man exposed to PCBs has not been established. However, the nausea and anorexia they experienced was suggestive of gastritis. Liver hypertrophy, proliferation of the hepatic endoplasmic reticulum and increased microsomal enzyme activity were consistent hepatic changes in both species (1, 11). Altered ketosteroid levels, dysmenorrhea, reduced birth weights of infants born to exposed mothers and transplacental movement of PCBs have also been recorded. Of major significance is the persistence of lesions in both species for years following exposure to PCBs which may be related in part to the continued presence of detectable levels of PCB in the tissues of those exposed (4, 19).

The previously presented data indicate that the PCBs are toxic to primates over a wide dose range. It is particularly significant that monkeys develop signs of PCB intoxication within 1 to 2 months at doses as low as 2.5 and 5.0 ppm in the diet. Although the possibility of man consuming a diet containing these concentrations of PCB on a continuous basis is remote, these data do point out the fact that only small amounts of these compounds are required to produce toxicity and that a safe level has not yet been established. Since PCBs are known to accumulate in the tissues of exposed animals, continuous exposure to even minute quantities may be sufficient to eventually cause toxic effects.

Indications are that the toxicity of PCB is directly related to the ability of the animal to metabolize and excrete it (3, and Allen and Van Miller, ms in preparation). This has been demonstrated in rats and monkeys by the use of drugs that inhibit and increase the activity of hepatic microsomal enzymes. Animals that had been pretreated with phenobarbital prior to PCB administration suffered no ill effects from doses of PCB that under control conditions would have been very toxic. However, when the animals were given SKF 525A during the initial 24 hours of exposure to PCB, they succumbed rapidly. Such data suggest that the PCBs and not their metabolites are responsible for the acute toxic effects of the PCBs.

It has recently been shown that rats were able to survive for a period of one year on doses of PCB (100 ppm in the diet) that were lethal to monkeys within 2 to 3 months (2, 6). An explanation for the variation in species response is associated with the rate at which they metabolize PCB. In the rat the half-life of the PCB isomer 2,5,2′,5′-tetrachlorobiphenyl was less than 24 hours (21) due to its rapid metabolism and excretion. However, the half-life of 2,5,2′,5′-tetrachlorobiphenyl in the monkey is much longer (Allen and Van Miller, ms in preparation), primarily due to its slow metabolism. This difference in the rate of metabolism of 2,5,2′,5′-tetrachlorobiphenyl in the rat and monkey may be related to the difference in metabolic pathways. There are data which indicate that most of the PCBs undergo direct hydroxylation in the rat (21) while an arene oxide intermediate is formed prior to hydroxylation in the monkey (12).

The presence of arene oxide as an intermediate metabolite of PCB in the primate broadens the scope of possible injurious effects that may be experienced following PCB exposure. The arene oxides have been shown to be strong alkylating agents capable of producing sufficient changes in the proteins and nucleic acids to cause death or neoplastic transformations in affected cells (18). These cell modifications may be rather subtle and require years to be manifested. Therefore, continuous low level exposure to PCBs may be insufficient to produce obvious signs of PCB intoxication yet the dose may be adequate to cause macromolecular alterations.

Many of the lesions that develop in PCB intoxicated primates are directly related to the deposition of the compound in the affected tissue while other changes may be secondary. The loss of weight, decrease in serum protein and moderate anemia were likely related to the reduced food intake that occurred in the animals with hyperplastic gastritis, the latter having developed due to the irritating effect that the PCB had on the gastric mucosa. The hypertrophic hyperactive livers that developed in PCB exposed animals were a result of the stimulatory effect they had on the hepatic endoplasmic reticulum and their enzymes. However, when the exposure to PCB is sufficiently great, degenerative changes develop particularly in the organelle containing the highest level of PCB, the endoplasmic reticulum (1). The skin lesions may also be directly related to the presence of high levels of PCB. It has been shown that following exposure a considerable amount of PCB accumulates in the skin and underlying tissue and persists in this area for an indefinite period (4).

Many of the injurious effects of PCB intoxication may result secondarily from the rapid metabolism of endogenous substances by the hyperactive endoplasmic reticulum. Following PCB exposure there is an increased metabolism of steroids and steroid-like compounds. These altered steroid hormone levels may be responsible for the inability of PCB-fed primates to conceive (8). Decreased levels of, vitamin A as have been reported to occur in quail, rats (9) and monkeys (D. A. Barsotti, M. H. Zile and Allen, unpublished observations) could also be at least in part responsible for the follicular dermatitis that occurs in these animals.

A majority of the data presented in this paper were obtained through the efforts of my colleagues in the Experimental Pathology Laboratory: L. J. Abrahamson, D. A. Barsotti, L. A. Carstens, I. C. Hsu, R. J. Marlar, D. H. Norback, J. P. Van Miller.

REFERENCES

1. ALLEN, J. R., L. J. ABRAHAMSON AND D. H. NORBACK. Biological effects of polychlorinated biphenyls and triphenyls on the subhuman primate. *Environ. Res.* 6: 344, 1973.
2. ALLEN, J. R., L. A. CARSTENS AND L. J. ABRAHAMSON. Responses of rats exposed to polychlorinated biphenyls for fifty-two weeks. I. Comparison of tissue levels of PCB and biological changes. *Arch. Environ. Contam. Toxicol.* In press.
3. ALLEN, J. R., L. A. CARSTENS, L. J. ABRAHAMSON AND R. J. MARLAR. Response of rats and nonhuman primates to 2,5,2',5'-tetrachlorobiphenyl. *Environ. Res.* 9: 265, 1975.
4. ALLEN, J. R., L. A. CARSTENS AND D. A. BARSOTTI. Residual effects of short-term, low level exposure of nonhuman primates to polychlorinated biphenyls. *Toxicol. Appl. Pharmacol.* 30: 440, 1974.
5. ALLEN, J. R., L. A. CARSTENS AND D. H. NORBACK. Biological effects of the polychlorinated biphenyls in nonhuman primates. *Proceedings of the International Symposium on Recent Advances in the Assessment of the Health Effects of Environmental Pollution*, World Health Organization. In press.
6. ALLEN, J. R., AND D. A. NORBACK. Polychlorinated biphenyl and triphenyl induced mucosal hyperplasia in primates. *Science* 179: 498, 1973.

7. ALLEN, J. R., D. H. NORBACK AND I. C. HSU. Tissue modifications in monkeys as related to absorption, distribution and excretion of polychlorinated biphenyls. *Arch. Environ. Contam. Toxicol.* 2: 86, 1974.

8. BARSOTTI, D. A., AND J. R. ALLEN. Effects of polychlorinated biphenyls on reproduction in the primate. *Federation Proc.* 34: 338, 1975.

9. BITMAN, J., H. S. CECIL AND S. J. HARRIS. Biological effects of polychlorinated biphenyls in rats and quail. *Environ. Health Persp.* 1: 145, 1972.

10. GOOD, C. K., AND N. PENSKY. Halowax acne; cutaneous eruptions in marine electricians due to certain chlorinated naphthalenes and diphenyls. *Arch. Dermatol. Syphilol.* 48: 215, 1943.

11. HIRAYAMA, C., T. IRISA AND T. YAMAMOTO. Fine structural changes of the liver in patients with chlorobiphenyl intoxication. *Fukuoka Igaku Zasshi* 60: 455, 1969.

12. HSU, I. C., J. P. VAN MILLER AND J. R. ALLEN. Metabolic fate of ^{3}H 2,5,2',5'-tetrachlorobiphenyl in nonhuman primates. *Bull. Environ. Contam. Toxicol.* 14: 233, 1975.

13. HUTTERER, F., F. SCHAFFNER, F. M. KLION AND H. POPPER. Hypertrophic, hypoactive smooth endoplasmic reticulum: A sensitive indicator of hepatotoxicity exemplified by dieldrin. *Science* 161: 1017, 1968.

14. JENSEN, S. Report on a new chemical hazard. *New Sci.* 32: 612, 1966.

15. JONES, J. W., AND H. S. ALDEN. An acneform dermatergosis. *Arch. Dermatol. Syphilol.* 33: 1022, 1936.

16. KOLBYE, A. C. Food exposure to polychlorinated biphenyls. *Environ. Health Persp.* 1: 85, 1972.

17. KURATSUNE, M. An epidemiologic study of "Yusho" or chlorobiphenyl poisoning. *Fukuoka Igaku Zasshi* 60: 513, 1969.

18. KURATSUNE, M., T. YOSHIMURA, J. MATSUZAKA AND A. YAMAGUCHI. Epidemiologic study on Yusho, a poisoning caused by ingestion of rice oil contaminated with a commercial brand of polychlorinated biphenyls. *Environ. Health. Persp.* 1: 119, 1972.

19. MILLER, J. A. Carcinogenesis by chemicals. *Cancer Res.* 30: 559, 1970.

20. OKIMURA, M., C. HIRAYAMA AND M. UZAWA. Study of Yusho (chlorobiphenyl poisoning) in clinical examination. *Fukuoka Igaku Zasshi* 62: 123, 1971.

21. VAN MILLER, J. P., I. C. HSU AND J. R. ALLEN. Distribution and metabolism of ^{3}H 2,5,2',5'-tetrachlorobiphenyl in rats. *Proc. Soc. Exptl. Biol. Med.* 148: 682, 1975.

22. YOBS, A. R. Food exposure to polychlorinated biphenyls. *Environ. Health Persp.* 1: 79, 1972.

Cyclical changes in the sexual skin of female rhesus: relationships to mating behavior and successful artificial insemination[1]

J. A. CZAJA, S. G. EISELE, AND R. W. GOY

*Wisconsin Regional Primate Research Center
and Department of Psychology*
University of Wisconsin, Madison, Wisconsin 53706

Estimating the time of ovulation or conception is an important aspect in much of the work on primate reproduction. Although there are increasingly successful and sophisticated techniques used in the laboratory to help predict ovulation, most are either expensive with respect to instrumentation or require expertise in physical examination of the animal. There are many situations, such as in the field, in group living cages, or in large scale breeding programs, where these techniques are less practical. Certain species of Old World primates show marked external changes in the size and color of tissues around the female perineum (sex skin) which have been related to ovarian condition (*Pan troglodytes* (21), *Papio hamadryas* (22), *Macaca nemestrina* (2), *Cercopithecus talapoin* (17)). However, there are, in comparison, only minimal changes of this type in certain of the primate species most widely used in research, such as the rhesus (*Macaca mulatta*).

The rhesus does show an edema and redness of the skin around the perineum, but these features are most pronounced in young females, the edema particularly becoming much less prominent in older and multiparous animals (4, 8). Although a degree of cyclicity in the changes of the rhesus sex skin across menstrual cycles has been noted, these changes have been dismissed by many as having little value in estimating or predicting ovulation (4, 8, 19). Nevertheless, our experience in mating large numbers of rhesus and our data on rhesus sex skin and color indicated systematic changes in color indexes

[1] Supported by grant RR-00167 from the National Institutes of Health and grant MH-21312 from the National Institute of Mental Health. Publication Number 14–027 of the Wisconsin Regional Primate Research Center.

83

during ovulatory menstrual cycles, even among older females. A study was therefore undertaken in which the external physiological signs were recorded and females artificially inseminated only once during each menstrual cycle. This procedure allowed *1*) empirical documentation of ovulation (i.e., pregnancy), *2*) evaluation of color measures as a gauge of ovarian condition and the time of maximum fertility, *3*) documentation of color changes across known ovulatory cycles and the pattern of these changes relative to day of successful insemination, and *4*) comparison of cycle day versus the status of the sexual skin and its color as a predictor of ovulation. In addition, a second study was undertaken to evaluate the relationship between changes in the sexual skin and patterns of mating behavior observed during conceptional cycles.

EXPERIMENT I

Materials and methods

Subjects: One hundred nine adult rhesus females from the Wisconsin Regional Primate Research Center Breeding Colony were artificially inseminated from one to six times during a 3-year period. Subjects were maintained under temperature-controlled conditions in individual cages. Females had free access to water and were fed a standard ration of Purina Monkey Chow once daily, usually between 7 and 8 AM.

Menstruation: The caged female rhesus were readily trained to turn their perinea toward the investigator. Aided with a flashlight, an investigator checked the cage pan and vaginal orifice of each female daily for signs of menstrual discharge and recorded all positive instances.

Sex skin color: The intensity and extent of color changes on adult rhesus females are quite individualistic. Color changes are seen around the perineum and on the thighs, on the abdomen, and on the face. All rhesus females do not reach the same color peak, so no absolute criteria can be established for estimating beforehand when this peak is reached. Females must consequently be followed daily through several menstrual cycles to establish baseline data on the extent and manner of these color changes for individual females. Females are scored by a general system of grading colors from 0 (virtually white) to 4.25 (very deep red) in steps of 0.25. The monkeys' thigh region is normally monitored for skin tone color and presence of edema. In most females the thigh area displays the greatest color change throughout the menstrual cycle, and it is easy to observe. Although nearly all females show good color changes in this area, there are certain females for which color cannot readily be scored on the thighs, and either the face or abdomen is used as an alternate site.

The color of females was checked between 10 AM and 1 PM daily while they were in their home cages. An important aspect of these determinations is that they be made under quiet and nonstressful conditions. The colored cutaneous areas will blanch when the female is stressed, thereby introducing unacceptable variability into color measurements. The most prominent and easily observable midcycle change in rhesus color is a relatively rapid and consistent decline or fading from peak sex skin color. We defined this change (color breakdown) as a successive decline in female color for 2 or more consecutive days. The first day on which such a color decrease is observed is labeled day of color breakdown and is noted as such on the female's record.

Sperm collection, evaluation, and selection: Semen was obtained from male rhesus monkeys with an electro-ejaculation technique modified from that described by Mastroianni and Manson (11). The first third of a monkey's ejaculate was collected in a centrifuge tube preheated to 39 C. The tube with the ejaculate was immediately placed into a 39 C incubator for 2 min and then removed and allowed to cool to room temperature for 10 min. At that time any coagulum was removed from the tube with sterile forceps. The number of sperm per milliliter was next determined by removing 20 μl of semen and in a two-step procedure, diluting it 50,000 to 1 with sterile pyrogen-free sodium chloride. This solution was run through a Coulter Counter Model B with a 100 μm aperture set for a sperm cell size of 2.5×5.0 μm.

A motility rating for the sperm cells was established by microscopic inspection of raw semen and a live-dead stain then used to establish the live-dead percentage. Semen with a rating of less than 50% live sperm was not used. Once all the semen parameters were established, the raw semen was diluted to about 200 million live sperm per ml. The diluent included 25% egg yolk extender, 0.20 M Tris buffered glucose without glycerine, streptomycin sulfate, penicillin G, and buffered potassium.

Insemination: Females were observed until their color index was sufficiently high relative to their previous cycles to indicate that they were close to peak coloration. At that point, females were removed from their cage for a single insemination with approximately 0.10 ml of the fresh diluted male semen collected about 0.5 hr earlier.

The female was first placed ventrally on a table and then a nasal speculum was used to spread the vaginal opening, exposing the cervix. Once any cervical mucin was removed, a 250 μl syringe with a blunted 2-inch, 18 gauge needle was threaded through the cervix and into the uterus for injection of the semen.

Results

A total of 218 artificial inseminations were performed, and of these 87, or 39.9%, produced pregnancies. Sixty-nine of the females became pregnant at least once, and 18 were successfully impregnated twice. The menstrual and color parameters for the 87 cycles with conception are shown in Fig. 1 aligned by day of successful insemination. The mean menstrual cycle day of insemination was calculated to be 12.0. This was followed 19.4 days later, on the average, by a period of postconceptional bleeding (pregnancy sign (8)). Color breakdown occurred an average 2.8 days after insemination as shown in the top graph of Fig. 1. It followed insemination in 99% of the cases, always occurred within 6 days of insemination, and never preceded insemination. The temporal relationship between successful inseminations and the day of color breakdown showed relatively small variability in comparison to their temporal relationships to the recorded periods of menstruation. Statistical analysis verified that the interval between the day of color breakdown and successful insemination was significantly less variable than that between insemination and the onset either of the menses preceding or of the pregnancy sign following it ($F = 3.97$, $P < 0.01$; $F = 7.00$, $P < 0.01$ respectively).

The bottom graph of Fig. 1 indicates the pattern of midcycle skin color changes relative to the mean periods of vaginal bleeding bounding the cycle of conception. Average color increased in intensity, reached a peak the day after successful insemination, and then dropped sharply over the succeeding 3 days. The mean duration of the pregnancy sign, 12.1 days, was much longer than that of the preceding menses, 2.9 days. However, the observable discharge of blood during the pregnancy sign was less constant and did not uniformly occur on every day throughout the duration of the pregnancy sign. While blood was always noted on the intervening days between the first and last days of the regular menses interval, it was observed only on the average of 72.6% of the days within the boundaries of the pregnancy sign.

When bred during limited portions of the menstrual cycles, rhesus monkeys have commonly been inseminated between day 11 and day 14 of the cycle, based on some long-standing data suggesting that this is the most fertile portion of the rhesus menstrual cycle (8, 20). In our study, where inseminations were timed relative to a period of intense skin coloration, successful inseminations occurred between day 7 and day 20 of the cycle relative to the onset of menstruation. Given that females met the physiological criterion of intense skin coloration, it then did not make any substantial difference on which day of the cycle they were inseminated. The bar graph in the left section of Fig. 2 shows the observed proportions of single inseminations that resulted in conception during the hypothesized

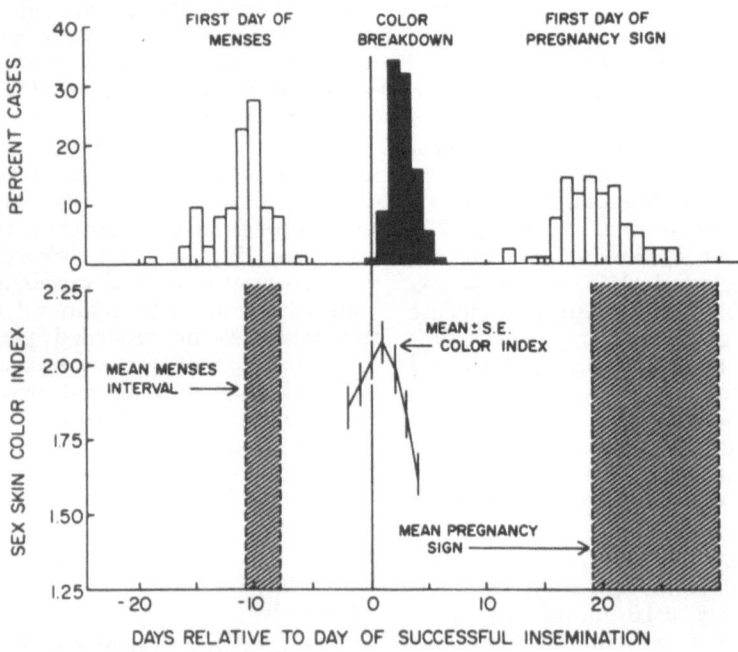

Figure 1. Physiological parameters of the female rhesus monkey relative to the day of successful insemination (no. = 87). *Top graph:* Initial day of vaginal bleeding prior to and after insemination plus the distribution of recorded sex skin color breakdown. *Bottom graph:* The pattern of sex skin color changes (Mean ± SE) around conception plus the mean menses and pregnancy sign intervals.

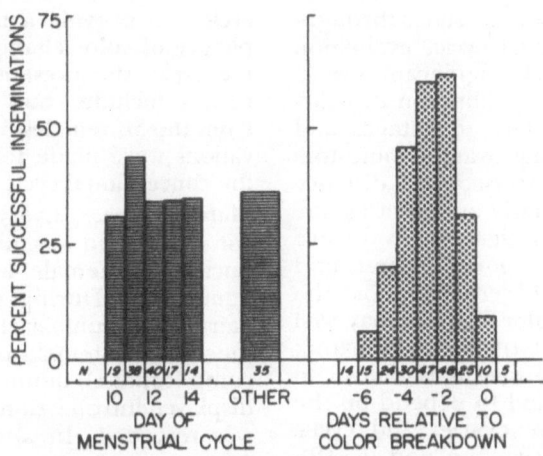

Figure 2. Success of single artificial insemination attempts during the menstrual cycle relative to day of the cycle or sex skin color breakdown (218 insemination attempts).

most fertile days of the cycle compared to the overall success ratio for inseminations on other days of the cycle, i.e., days 7 through 9 and 15 through 20 combined. Although the percentage of successful inseminations was greatest on day 11 (44.7%) there were no distinct differences between the supposedly most fertile days and others. The particular day of the cycle was therefore not in itself a good discriminator of insemination success. However, there was a strong relationship between insemination success and color breakdown. The right hand portion of Fig. 2 depicts the probability that inseminations were successful as a function of the time of occurrence of color breakdown. Inseminations falling 2 to 4 days prior to the occurrence of color breakdown produced an average conception rate of 57.6%. The highest conception rate, 62.5%, occurred 2 days prior to color breakdown. This success ratio dropped to less than 10% for inseminations carried out either 6 days prior to breakdown or on the day of breakdown itself. Of 14 inseminations carried out 7 days prior to breakdown, as for 5 carried out 1 day after breakdown, none were successful.

The seasonality in gonadal function reported for rhesus under free-ranging conditions (3, 10, 16) also continues to a lesser degree in the laboratory (13–15). Maximum fertility of laboratory rhesus in the Northern Hemisphere reportedly occurs between August and January (19). In our study, inseminations were performed in every month of the year, but there was no evidence of seasonality in the ability of the single inseminations to produce conceptions. The overall success ratio between August and January was 40.9%, while that between February and July was 38.4%, and the two samples were well balanced with respect to the time of insemination relative to color breakdown.

EXPERIMENT II

The results reported above suggest a close relationship between ovulation, periovular hormonal changes, and observable changes in the status and color of the sexual skin. Given the varying endocrine states throughout the menstrual cycle, evaluation of reproductively significant events often requires equilibration of ovarian condition. The advantages and utility of external ovarian indicators which can be assessed at a distance become particularly apparent as one attempts to investigate reproductive interactions in more natural and complex social settings. For the rhesus, cyclic color changes may well provide this tool. Inasmuch as mating behavior of the rhesus is known to vary cyclically and to depend on the presence of the ovaries and cyclic changes in ovarian function (5, 12), it follows that some definable relationship should exist between mating activity and color of the sexual skin. Data substantiating this hypothesis for laboratory mating situations were obtained in the following studies, and they provide both an independent validation of sexual skin measures as an indicator of ovarian events as well as an example of its potential in research on primate reproduction.

Materials and methods

Subjects: Each of 57 adult rhesus females from the breeding colony was paired during a conceptional cycle with one of 15 proven breeding males. Owing to the circumstance that the number of males was limited, only 27 of these females were paired with a male on a daily basis throughout the cycle of conception. The remainder were paired a variable number of times ranging from 5 to 16. Only conceptional cycles were used for this analysis in order to eliminate atypical patterns associated with anovulatory cycles, abnormal luteal phases, and other irregularities.

Sex skin color: Ratings were made as described for **Experiment I**. For certain of the females, data were limited to the midportion of the menstrual cycle. To provide a more extended picture of color changes throughout the cycle, the presentation of color results includes only those ratings from the 36 females on which observations were made for every day of the conceptional cycle.

Mating behavior: Every pairing lasted for 12 min and was initiated by introducing the female into the male's home cage. During this period of pairing, the number of times the male touched the female on the back or rump (contact), mounted the female, displayed intromissions, or ejaculated was recorded. In addition, records were made of every sexual presentation by the female that occurred immediately after a contact by the male as well as every presentation that was maintained during contact and mounting by the male. The sum of these two types of presentations divided by the total number of contacts provides a ratio score that defines the *Presentation Quotient* (PQ). This ratio varies between the limits of 0 and 1.00 and higher scores reflect a stronger tendency of the female to respond to the male's touch by presentation. Pairings in which the male never contacted the female, i.e., in which the denominator would be zero, were not used in the calculation of averages.

Results

Both the incidence of ejaculation by the male and Presentation Quotient of the female were found to vary systematically with the sex skin color index throughout the cycle of conception (Fig. 3). Although the behavioral measures not unexpectedly were quite variable from day to day, both the percent of pairings on which the male ejaculated and the PQ were low at the beginning of the cycle and increased more or less regularly to peak values 2 days prior to color

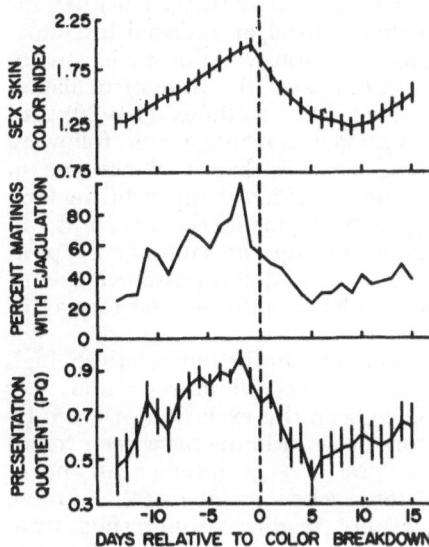

Figure 3. Changes in mating behavior and sex skin color throughout the cycle of conception. *Top graph*: Mean ± SE of sex skin color ratings estimated daily for 36 females. *Middle graph*: percent of pairings during the cycle on which the male ejaculated (no. = 57). *Bottom graph*: proportion of times female responded to male contact by presenting (see METHODS: Mean ± SE, no. = 57).

breakdown. From that time forward through the remainder of the cycles, both behavioral measures decreased steadily, reaching their nadirs 5 days after breakdown. Succeeding values tended to increase, and 15 days after breakdown, close to the onset of the pregnancy sign for the majority of females (see Fig. 1), the behavioral measures and the sex skin color index showed noticeable increases.

The peak values for behavior and for the sex skin color index were not coincident. Whereas the behavioral measures showed maximum values 2 days prior to breakdown, the color changes did not show maximum values until 1 day later, that is 1 day prior to breakdown.

The results of these natural inseminations are in accord with the results obtained with the artificial inseminations. All of these females that became pregnant were paired with males that ejaculated on days 4, 3, or 2 prior to breakdown. Inasmuch as pairings were not limited to a single day of the cycle, however, the data do not allow estimations of the optimal day for mating.

DISCUSSION

The results of these studies indicate that there are readily observable external physiological features on adult female rhesus monkeys which correlate closely with the period of maximum fertility during the menstrual cycle and with the mating performance of heterosexual pairs of rhesus. An index reflecting skin coloration showed systematic changes around the day of insemination for those cases resulting in pregnancy.

The most easily discernible color change was the relatively abrupt decrease from peak coloration, defined as color breakdown. Although the manner and degree of such color changes are individualistic, there appears to be sufficient consistency within females to offer the potential to estimate ovulation with a fair degree of reliability. Utilizing skin color as a guide for the time of artificial insemination, relatively high rates of conception were achieved with only single inseminations during each menstrual cycle.

Inseminations most successful in producing pregnancies occurred 2 to 4 days prior to color breakdown. This correlation suggests a strong relationship between skin color and periovular hormone changes. Average peak color and incidents of color breakdown occurred after the time of successful insemination and are therefore most useful in retrospective analyses. Nevertheless, the interval from day of successful insemination to color breakdown was significantly less variable than the period separating this day from menses. In lieu of more complicated and expensive analyses of ovarian condition, knowing the day of color breakdown allows one to more closely estimate the period of maximum fertility and, within days of insemination, predict the probability that an insemination would prove successful. We have, unfortunately, not been able to discern any color parameters that can positively identify that ovulation is imminent. Absolute peak color varies among females and to a lesser degree for the same female from cycle to cycle. In certain cycles, females are also found to maintain maximum color for 2 to 4 days. Thus, intense coloration per se can be used only to identify an interval of relatively high fertility. Despite this shortcoming, one potential value of using color in studies of rhesus reproduction is amply demonstrated by the high pregnancy rate achieved in this study.

Hartman, in his classic report (8), noted a 16.1% rate of conception with single matings during the portion of the cycle he thought to be the most fertile (days 11 to 14) and a conception rate with natural inseminations of 11.2% on other days of the cycle (days 3 to 10 and 15 to 28 inclusive). Calendar methods such as this are usually more effective when cycles are consistent in length. Rhesus menstrual cycles, however, are not known for such constancy. This is a major reason why there are advantages to a method that more directly reflects ovarian condition. Overall, 39.9% of our inseminations resulted in pregnancies and 48.2% of all the females inseminated became pregnant with the first insemination. Of the females who did get pregnant with our methods, 76.5% conceived on the first insemination attempt. This proportion of successful inseminations is higher than rates for either natural or artificial inseminations previously reported in studies that estimated the time of ovulation by calendar methods (18–20). An additional advantage of following skin color is that it allows one to monitor ovarian changes during long periods of amenorrhea, such as can occur during the summer or postpartum, and detect possible incidents of ovulation prior to the display of menses.

One reason for our relatively high rate of successful impregnations may have been the exclusion of anovulatory cycles. Rhesus have been found to have a large number of anovulatory cycles in captivity (4) and these are not readily distinguished from ovulatory cycles on the basis of cycle length or menstrual parameters (8). In the studies reported here, we inseminated only during cycles with color fluctuations. We are now finding that a large class of anovulatory cycles can readily be distinguished by their lack of color changes across the cycle. However, significant color changes or a clear color breakdown do not appear sufficient to identify fertile cycles, since a substantial proportion of cycles showing these changes do not produce pregnancies despite repeated inseminations by either natural or artificial means.

Changes in skin color also were found to serve as a reliable guide for predicting the time in the cycle when pairing the female with a male will most likely result in effective copulation within a short period. In this regard it should be noted that although the male's ejaculatory performance is closely associated with the female's changes in color, color per se does not determine the male's behavior. Color is most intense the day following the greatest likelihood of ejaculation during repeated daily time-limited pairings. It seems most likely that both skin color and specific components of rhesus mating behavior are influenced by the same closely correlated endocrine factors associated with ovulation (1, 9). As such, it is important to note that primates like the rhesus retain the adaptive processes that help to insure impregnation by maximizing the probability of copulation to ejaculation during the most fertile portion of ovulatory menstrual cycles.

The high rate of conception we found possible with an artificial insemination procedure involving semen dilution is particularly relevant in light of the increasing dependence on laboratory breeding as a source of primates for research. The methods outlined allow rhesus to be used throughout the year in an efficient manner. It is also becoming increasingly obvious that certain laboratory rearing conditions can have a negative influence on the development of adequate primate sexual behavior (6, 7). Artificial insemination is one of the few avenues of recourse for breeding animals that were raised lacking the social environment necessary for normal adult sexual performance.

REFERENCES

1. ALLEN, E. The menstrual cycle in the monkey *Macacus rhesus*: Observations on normal animals, the effects of removal of the ovaries and the effects of injections of ovarian and placental extracts into the spayed animal. *Carnegie Inst. Publ. Contrib. Embryol.* 19: 1–43, 1927.
2. BULLOCK, D. W., C. A. PARIS AND R. W. GOY. Sexual behavior, swelling of the sex skin, and plasma progesterone in the pigtail macaque. *J. Reprod. Fertility* 31: 225–236, 1972.
3. CONAWAY, C. H., AND D. S. SADE. The seasonal spermatogenic cycle in free-ranging rhesus monkeys. *Folia Primatol.* 3: 1–12, 1965.
4. CORNER, G. W. Ovulation and menstruation in *Macacus rhesus*. *Carnegie Inst.*

Publ. Contrib. Embryol. 15: 73–101, 1923.
5. GOY, R. W., AND J. RESKO. Gonadal hormones and behavior of normal and pseudohermaphroditic nonhuman female primates. *Recent Progr. Hormone Res.* 28: 707–733, 1972.
6. GOY, R. W., K. WALLEN AND D. A. GOLDFOOT. Social factors affecting the development of mounting behavior in male rhesus monkeys. In: *Reproductive Behavior*, edited by W. Montagna and W. A. Sadler. New York: Plenum, 1974, p. 223–247.
7. HARLOW, H. F. Sexual behavior in the rhesus monkey. In: *Sex and Behavior*, edited by F. A. Beach. New York: Wiley, 1965, p. 234–265.
8. HARTMAN, C. G. Studies in the reproduction of the monkey macacus (*Pithecus*) rhesus, with special reference to menstruation and pregnancy. *Carnegie Inst. Publ. Contrib. Embryol.* 23: 1–161, 1932.
9. HOTCHKISS, J., L. E. ATKINSON AND E. KNOBIL. Time course of serum estrogen and luteinizing hormone (LH) concentrations during the menstrual cycle of the rhesus monkey. *Endocrinology* 89: 177–183, 1971.
10. KOFORD, C. B. Population dynamics of rhesus monkeys on Cayo Santiago. In: *Primate Behavior*, edited by I. Devore. New York: Holt, Rinehart and Winston, 1965, p. 160–174.
11. MASTROIANNI, L., AND W. A. MANSON. Collection of monkey semen by electro-ejaculation. *Proc. Soc. Exptl. Biol. Med.* 112: 1025–1027, 1963.
12. MICHAEL, R. P., J. HERBERT AND J. WELEGALLA. Ovarian hormones and the sexual behavior of the male rhesus monkey (*Macaca mulatta*) under laboratory conditions. *J. Endocrinol.* 39: 309–310, 1967.
13. PLANT, T. M., D. ZUMPE, M. SAULS AND R. P. MICHAEL. An annual rhythm in the plasma testosterone of adult male rhesus monkeys maintained in the laboratory. *J. Endocrinol.* 62: 403–404, 1974.
14. RIESEN, J. W., R. K. MEYER AND R. C. WOLF. The effect of season on occurrence of ovulation in the rhesus monkey. *Biol. Reprod.* 5: 111–114, 1971.
15. ROBINSON, J. A., AND W. E. BRIDSON. Sexual behavior and plasma androgen concentration in laboratory-housed rhesus males: Effects of age and season. *Soc. Study Reprod.*, 7th Ann. Meeting, Ottawa, Canada, Abstract #75, 1974.
16. SADE, D. S. Seasonal cycle in size of testes of free-ranging *Macaca mulatta*. *Folia Primatol.* 2: 171–180, 1964.

17. SCRUTON, D. M., AND J. HERBERT. The menstrual cycle and its effect on behaviour in the Talapoin monkey (*Miopithecus talapoin*). *J. Zool.* 162: 419–436, 1970.

18. SETTLAGE, D. S. F., S. SWAN AND A. G. HENDRICKX. Comparison of artificial insemination with natural mating technique in rhesus monkeys, *Macaca mulatta. J. Reprod. Fertility* 32: 129–132, 1973.

19. VALERIO, D. A., A. J. PALLOTTA AND K. D. COURTNEY. Experiences in large-scale breeding of simians for medical experimentation. *Ann. N.Y. Acad. Sci.* 162: 282–296, 1969.

20. VAN WAGENEN, G. Optimal mating time for pregnancy in the monkey. *Endocrinology* 37: 307–312, 1945.

21. YOUNG, W. C., AND W. D. ORBISON. Changes in selected features of behaviors in pairs of oppositely sexed chimpanzees during the sexual cycle and after ovariectomy. *J. Comp. Psych.* 37: 107–143, 1944.

22. ZUCKERMAN, S. AND A. S. PARKES. Observations on secondary sexual characters in monkeys. *J. Endocrinol.* 1: 403–439, 1939.

Endocrine and metabolic responses to cold in baboons[1]

C. C. GALE

Regional Primate Research Center
University of Washington, Seattle, Washington 98195

When mammals are stressed by cold, the endocrine system is stimulated leading to increased "chemical" thermogenesis and a rise in resting metabolic rate. Probably most fundamental to this response is increased activity of the sympathicoadrenomedullary system, and/or augmented sensitivity to the calorigenic action of catecholamines, especially norepinephrine. Thus, when cold stimuli are applied peripherally (e.g., exposure to a cold environment), or internally (e.g., cooling the hypothalamus, spinal cord, or abdomen), the sympathicoadrenomedullary system is activated in all mammalian species studied, including rat, pig, goat, sheep, rhesus monkey, baboon, and man (3). Increased sensitivity to the calorigenic action of exogenous norepinephrine is an accepted criterion for acclimation to cold, but may be valid only for mammals which develop brown adipose tissue, primarily rodents and hibernating animals. Usually, cold stress also stimulates secretion of adrenal cortical and thyroid hormones. However, acute cold stress may activate the adrenal cortex but not the thyroid, or vice versa (3). The underlying mechanism which preferentially mobilizes adrenal glucocorticoids versus thyroxine is unclear but likely relates to inhibitory actions of peripherally or centrally released hormones on the hypothalamic-pituitary axis. Because of variability in cold defense mechanisms among lower mammalian species, we have chosen to study an infrahuman primate phylogenetically close to man, the baboon (*Papio anubis*), as an experimental model of human thermoregulation. Prepuberal male baboons (11 to 19 kg) were adapted to primate chair restraint and then either exposed to 6 C ambient temperature (T_a) for 11 to 13 weeks, or subjected to local cooling and warming of the preoptic anterior hypothalamus for 1 to 3 hr in 22–25 C T_a. Certain animals were surgically thyroidectomized, and others were trained to bar press for infrared heat reward as an index of behavioral thermoregulation.

METABOLIC RESPONSE TO COLD

When baboons acclimated to 23–25 C T_a were exposed for 11 to 13 weeks

[1] Supported in part by National Institutes of Health research grants NS 06622 and RR 0166.

to 6 C T_a, their peripheral vasculature became persistently constricted but shivering, as monitored on closed-circuit television, was slight and sporadic and disappeared after several days. To evaluate alterations in metabolism during and after acclimation to cold, O_2 consumption was measured by an open system from 9 AM to 12 N. Resting metabolic rate was calculated during 5 to 15 min periods during which baboons remained motionless with eyes closed (apparent sleep). Calculations were not made during intervals of sleep associated with rapid fluttering of eyelids, presumably REM sleep, because the respiratory rate declined sporadically resulting in artifacts of O_2 consumption. Figure 1 shows that the resting metabolic rate was elevated approximately 20 to 30% during the 13 weeks in the cold. Baboons appeared to sleep less and be more restless then. The rise in metabolism was largely, if not entirely, via an increase in nonshivering thermogenesis. To elucidate whether the baboons acclimated to chronic cold by developing increased sensitivity to the calori-

genic action of norepinephrine, as occurs in rodents (3), the catecholamine was injected intramuscularly (250 μg/kg as bitartrate). The injection was given 1 hr after the chamber temperature had been raised to 24 C to allow the elevated resting metabolic rate to subside to precold control level. The fact that the resting metabolic rate declined in this time interval showed that thermoneutral metabolic rate was not raised in cold-acclimated baboons. Figure 2 shows that the augmentation in resting O_2 consumption by norepinephrine was only slightly greater than before cold exposure. Hence, increased sensitivity to norepinephrine did not appear to be a major factor in acclimation to cold. However, recent evidence suggests that in addition to marked rise in urinary norepinephrine (2), significant norepinephrine sensitivity does develop in cold-acclimated rhesus monkeys; administration of norepinephrine (250 μg/kg of bitartrate) led to a 40% rise in resting metabolic rate (1). These monkeys at sacrifice were found to possess yellow adipose tissue, a specialized

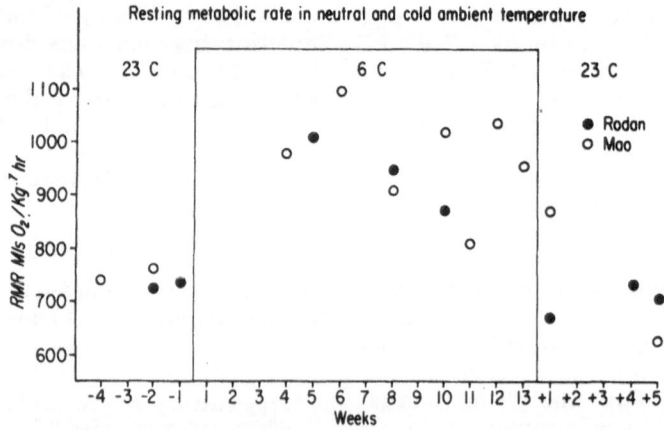

Figure 1. Elevation of resting metabolic rate (RMR) in 2 baboons kept 13 weeks at 6 C ambient temperature.

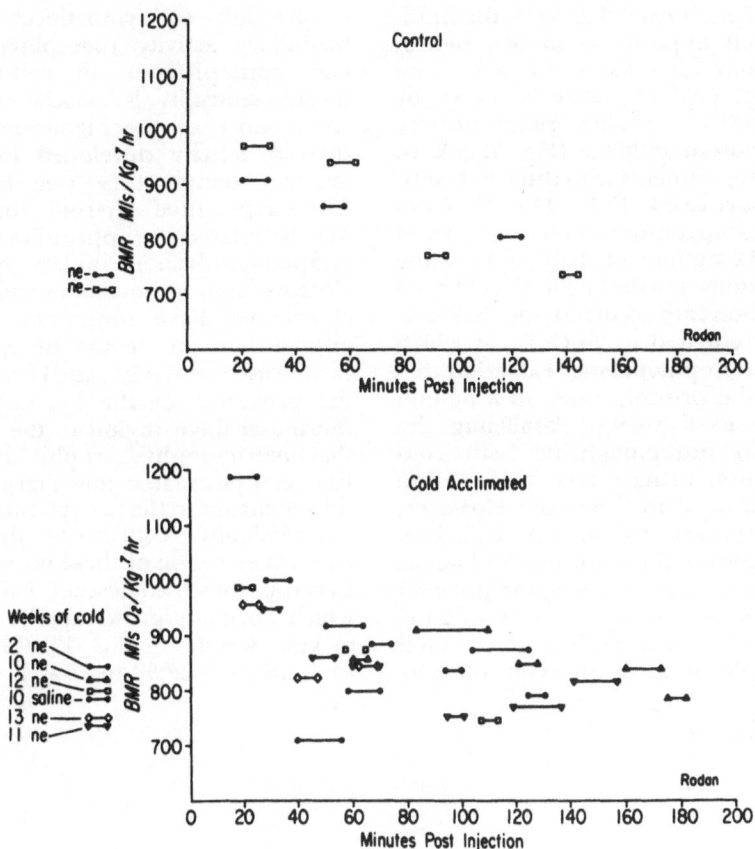

Figure 2. Increase in metabolic rate following intramuscular injection of norepinephrine (ne) (250 µg/kg as bitartrate) in a baboon before (upper diagram) and during cold-acclimation (lower diagram).

tissue that may serve a role similar to brown adipose tissue. Brown adipose tissue is found in cold-acclimated rodents, hibernators, and all mammalian neonates, and is extremely sensitive to the thermogenic action of norepinephrine. By its anatomic distribution, brown adipose tissue functions to provide heat to the spinal cord, heart, and kidney. However, its existence has not been documented in adult humans. The marginal rise in resting metabolic rate in cold-acclimated baboons by exog-

enous norepinephrine is consistent with the report of only 8 to 14% elevation in cold-acclimated humans given norepinephrine (10, 11). It is unknown whether cold-acclimated baboons develop yellow adipose tissue as do rhesus monkeys.

SYMPATHICO-ADRENOMEDULLARY RESPONSE TO COLD

Although increased sensitivity to norepinephrine was at best marginal in cold-acclimated baboons, the fundamental importance of this neurohormone for resistance to acute and chronic cold exposure is shown by the marked, sustained rise in urinary free norepinephrine (Fig. 3) (4). In four experiments with three baboons, exposure to 6 C T_a for 11 to 13 weeks evoked a prompt threefold rise, from 4 to 12 ng/min in daily 24-hr urine collections. No decline in the elevated excretion rate occurred until baboons were returned to 24 C T_a, at which time norepinephrine excretion fell *below* the pre-cold value to 2 ng/min for 5 to 6 weeks. Paralleling the rise in norepinephrine with cold exposure, urinary free epinephrine rose from 2 to 7 ng/min. However, in contrast to norepinephrine, epinephrine fell within several weeks to values intermediate to the pre-cold control. Returning baboons to 24 C T_a led to rapid decline of excreted epinephrine *below* the control. The suppression of sympathicoadrenomedullary activity (norepinephrine and epinephrine) on return to thermoneutrality is associated with continuance of the higher level of thyroid activity developed in cold ambient temperature (see below). This augmented thyroid function may be related to suppression of the sympathicoadrenomedullary system. Because thyroid hormones and catecholamines have important synergistic actions on metabolism and the cardiovascular system, and because of the proximity of the hypothalamic thermosensitive region to the hypothalamic hypophysiotrophic area, it has been postulated that thermosensitive neurons in the preoptic anterior hypothalamus coordinate the reciprocal secretion of these hormones. Because these adolescent baboons, which normally grow rapidly, failed to gain weight in 6 C T_a although food intake rose 50 to 75%, it would

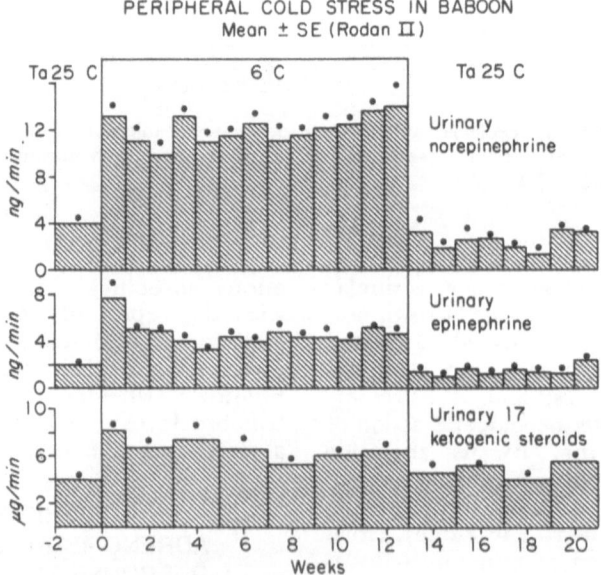

Figure 3. Rise in urinary norepinephrine, epinephrine, and 17-ketogenic steroids during chronic cold exposure in a baboon.

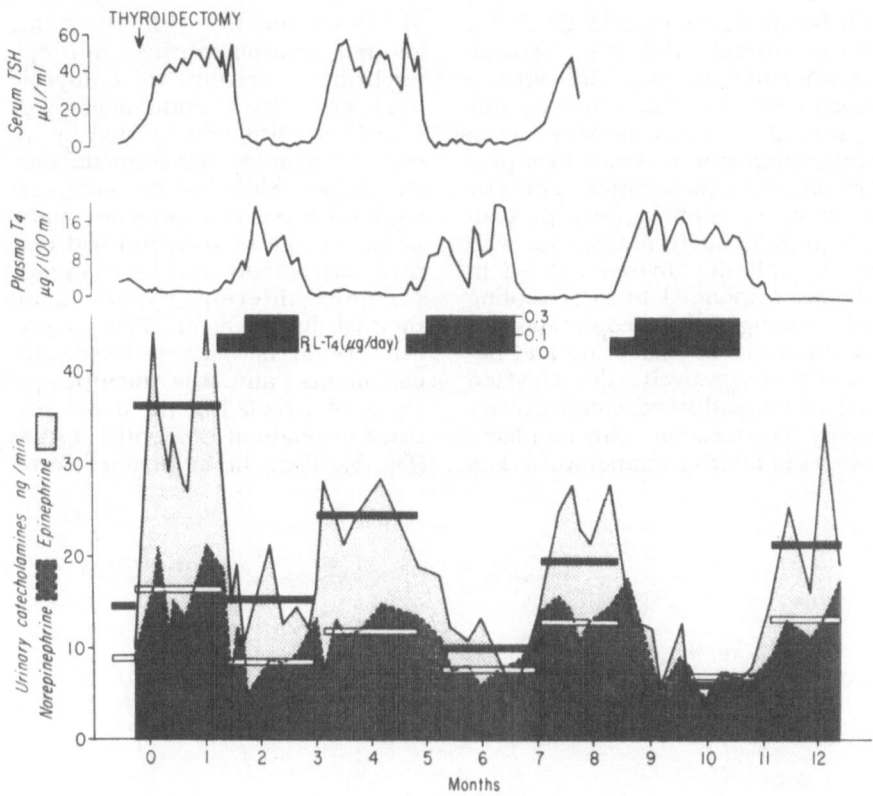

Figure 4. Increase in urinary norepinephrine and epinephrine, and in serum thyroid stimulating hormone (TSH) following surgical thyroidectomy in a baboon. Replacement with thyroxine (T4) lowered urinary catecholamines to or below preoperative level.

be expected that a period of protein anabolism ("catch-up growth") would follow cold exposure in which thyroid hormones would be essential. Parenthetically, the role of testosterone in protein anabolism of cold stressed adolescent baboons remains to be elucidated, but in a similar study in adult male rhesus monkeys, plasma testosterone was chronically lowered in the cold (2). Catecholamine-thyroid hormone interaction may not be the only factor mediating the reduction of sympathicoadrenomedullary activity below pre-cold control level. Residual tissue sensitivity to the calorigenic

effect of norepinephrine, as well as altered perception of thermoneutral T_a as a warm environment (i.e., increased gain of warmth detectors), may be contributory factors.

To study reciprocal control of thyroid and sympathicoadrenomedullary hormones as a function of thermosensitivity of the hypothalamus, four baboons were surgically thyroidectomized and subjected to local cooling and warming of the preoptic anterior hypothalamus. Figure 4 shows that following thyroidectomy urinary norepinephrine and epinephrine rose markedly in baboons maintained in 24 C T_a. These animals did not become hypothermic despite the virtual absence of thyroxine, presumably because of the compensatory rise in catecholamines to maintain heat production and conservation. The elevation of epinephrine was particularly prominent. In accordance with our hypothesis, thyroxine-deficient baboons responded to local cooling and warming of the preoptic anterior hypothalamus by increasing and decreasing, respectively, the elevated level of sympathicoadrenomedullary activity in association with regulated changes in internal temperature (Fig.

5). Replacement with thyroxine lowered norepinephrine and epinephrine excretion to euthyroid levels, and when a condition of severe hyperthyroidism was created by excessive thyroxine replacement, catecholamines fell below the preoperative level. But in all conditions, appropriate change in sympathicoadrenomedullary activity was elicited by local preoptic anterior hypothalamus thermal displacement. This reciprocal relation between secreted catecholamines and the manipulated levels of circulating thyroxine persisted throughout 12 months of study (Fig. 3). That the augmented secre-

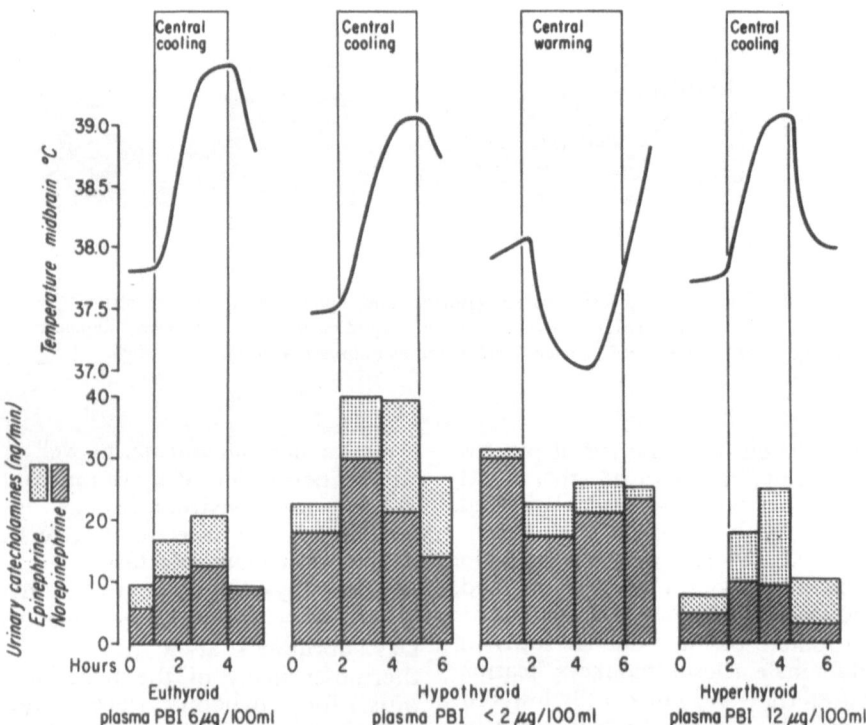

Figure 5. Local cooling of the preoptic anterior hypothalamus raised internal (midbrain) temperature in association with increased urinary catecholamines in a baboon. Surgical thyroidectomy caused severe hypothyroidism, and replacement with thyroxine led to severe hyperthyroidism. Preoptic anterior hypothalamus warming suppressed urinary catecholamines in the hypothyroid baboon in conjunction with a decline in internal temperature.

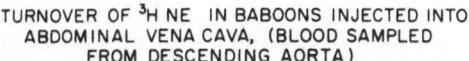

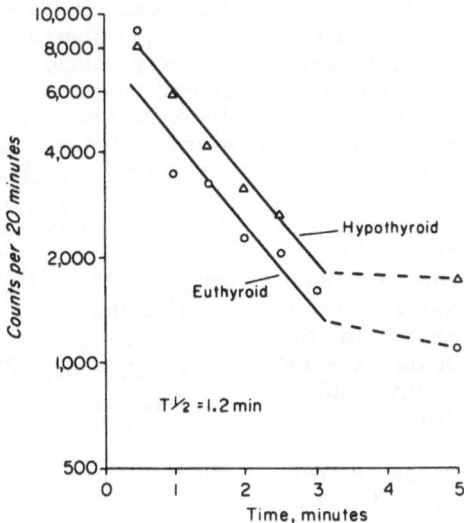

Figure 6. Turnover rate for [³H] norepinephrine (³H NE) in euthyroid and hypothyroid baboons.

tion of catecholamines in hypothyroidism was not secondary to more rapid clearance from the blood was shown by an identical circulatory half-life for norepinephrine, 1.2 min, in both hypothyroid and euthyroid baboons (Fig. 6). In this experiment, [³H]norepinephrine was injected intravenously and blood samples were drawn from conscious animals via an in-dwelling arterial catheter (Gale, Johnson, and Henry, unpublished).

ADRENOCORTICAL RESPONSE TO COLD

Exposure to 6 C T_a provoked a sustained twofold increase in secretion of glucocorticoids, as indicated by urinary 17-ketogenic steroids (Fig. 3) (4). Aside from general stress-coping actions, the augmented glucocorticoids in cold-acclimated baboons, which were fed twice daily, may be related to increased requirements for gluconeogenesis and lipid substrate mobilization, as well as to maintain the responsiveness of peripheral arterioles to the pressor action of catecholamines.

THYROID RESPONSE TO COLD

In smaller mammals, such as rodents, acute cold exposure elicits a prompt activation of the hypothalamic-pituitary-thyroid axis, evidenced by increased secretion of thyroid stimulating hormone and thyroid hormones, thyroxine and triiodothyronine (3). However, thyroidal activation by acute cold has not been demonstrated in human or infrahuman primates other than neonates (see below). To study thyroidal activity during prolonged cold exposure in baboons, the animals were injected i.v. with [¹²⁵I]thyroxine and [¹³¹I]triiodothyronine before, during, and after cold stress. Metabolic clearance rates for the thyroid hormones were calculated from the circulatory half-life and the distribution volume (Table 1). Production rates were determined from metabolic clearance rates and serum concentrations of the hormones. Relative to pre-cold control values, the metabolic clearance rate and production rate of thyroxine rose 20 to 30% within 2 weeks in the cold, and remained elevated up to 7 weeks after baboons had been returned to 24 C T_a. Because baboons in the cold consumed 50 to 75% more of the Purina monkey biscuit diet, the increased production of thyroxine is partially related to enhanced enterohepatic clearance and greater fecal excretion of nonmetabolized thyroxine (measured as fecal ^{125}I). However, feeding the baboons a high caloric, low residue synthetic diet while in the cold did not markedly lower thyroxine production rate although it reduced enterohepatic loss. Thus, baboon R (Table 1) when fed the low residue diet after the second week in the cold lowered its stool: urine ratio for ^{125}I from 0.5 to 0.3 (urinary ^{125}I being used as an index of metabolized thyroxine). During this interval, thyroxine production rate was only marginally reduced from 23 to 22 μg/day. Hence, these preliminary data suggest there is a greater metabolic requirement for thyroxine in cold-acclimated baboons. Probably of greater significance, the production rate of triiodothyronine rose progressively in the cold to 30 to 50% of control values (Table 1), and remained elevated at an intermediate level during the post-cold period. Since enterohepatic clearance has less influence on triiodothyronine metabolism than on thyroxine metabolism, the enhanced production provides strong evidence for augmented tissue requirements for tri-

iodothyronine in cold-acclimation. Current belief holds that most of the metabolic action of the thyroid is mediated through triiodothyronine converted peripherally from thyroxine. In concordance, when a surgically thyroidectomized baboon was given thyroxine replacement, serum levels of triiodothyronine were restored to euthyroid levels (Wu and Gale, unpublished). Thus, the elevation in thyroxine production in the cold may represent the provision of a precursor to the active hormone, triiodothyronine.

COOLING AND WARMING OF THE PREOPTIC ANTERIOR HYPOTHALAMUS

The patterns of endocrine and metabolic responses evoked by cold exposure in baboons are in general similar to those elicited by local cooling of the preoptic anterior hypothalamus, except that thyroidal activation does not occur in the latter case. Preoptic anterior hypothalamus cooling demonstrates the capability of thermodetector neurons in the thermoregulatory center to coordinate somatic, autonomic, and endocrine motor outflow to maintain thermal homeostasis. Cooling of the preoptic anterior hypothalamus for 1 to 2 hr in baboons elevates O_2 consumption about 50% in conjunction with cutaneous vasoconstriction and shivering. Urinary norepinephrine and epinephrine and plasma cortisol levels rise (6). But neither plasma thyroid stimulating hormone nor thyroxine increase (Gale and Webster, unpublished). Because preoptic anterior hypothalamus cooling "opens the loop" of thermal negative feedback, heat gain activities increase relative to heat loss, and the regulated level of internal temperature (T_{int}) is displaced upward about 0.5 to 1 C. This situation is analogous

TABLE 1. Metabolism of thyroxine and triiodothyronine in baboons in chronic cold

	Pre-cold	In cold			Post-cold	
		2 weeks	6 weeks	11 weeks	2 weeks	7 weeks
Thyroxine						
$T_{\frac{1}{2}}$, days	2.3	2.1	2.4		2.4	2.1
MCR,[a] l/day	0.56	0.73	0.60		0.59	0.72
Hormone level, μg/dl	3.1	3.2	3.7		3.8	3.5
Production rate, μg/day	17.0	23.0	22.0		22.0	25.0
Excretion ratio, Stool/urine	0.4	0.5	0.3		0.3	0.3
Triiodothyronine						
$T_{\frac{1}{2}}$, days	0.70	0.83	0.76	0.81	0.89	0.65
MCR,[a] l/day	12.0	12.2	11.2	11.4	15.0	15.5
Hormone level, ng/dl	120.0	150.0	175.0	194.0	114.0	118.0
Production rate, μg/day	14.3	18.4	19.6	22.1	17.1	18.3
Excretion ratio ·Stool/urine	0.1	0.2	0.1	0.2	0.1	0.1

[a] Metabolic clearance rate.

to pyrogenic fever. Preoptic anterior hypothalamus cooling provokes thermoregulatory behavior, as does pyrogenic fever (7): baboons which have been trained to bar press for infrared heat reward in 10 C T_a, markedly increase bar-pressing for warmth when subjected to preoptic anterior hypothalamus cooling in 24 C T_a. Similarly, lipid substrate is utilized more rapidly in overnight fasted baboons during preoptic anterior hypothalamus cooling in 24 C T_a (8), as indicated by accelerated turnover of isotopically labeled free fatty acids infused i.v. A further manifestation of the augmented sympathicoadrenomedullary activity provoked by preoptic anterior hy-

pothalamus cooling is the increase in regulated level of blood pressure and heart rate (12). This cardiovascular response reveals an interaction between hypothalamic warmth detectors and neurons in the preoptic anterior hypothalamus region mediating the baroreceptor reflex (Fig. 7). The capacity of preoptic anterior hypothalamus warmth receptors to integrate the cardiovascular system in thermoregulation is also shown by microinjection of prostaglandin E_1 into the preoptic anterior hypothalamus (Fig. 7). Central injection of this fatty acid derivative evokes a regulated rise in T_{int} ("fever") in all mammals, possibly by blocking the responsiveness of central warmth detectors to norepinephrine, the neurotransmitter providing thermal feedback information from the body core. Thus, injection of prostaglandin E_1 (1 μg) into the preoptic anterior hypothalamus of baboons led to a 1.3 C rise in T_{int} (Fig. 8). However, because of thermal inertia, this response required 1 to 2 hr to peak. By comparison, blood pressure and heart rate rose promptly within several minutes (there being no inertia) and then declined gradually as T_{int} rose. Of interest, this reciprocal relation between the evoked thermoregulatory response (rise in blood pressure) and the delayed achievement of thermal equilibrium (elevated T_{int}), is also seen for increased O_2 consumption and thermoregulatory behavior. That is, elevated metabolic rate and bar-pressing for warmth during preoptic anterior hypothalamus cooling occurs only during

the 1 to 2 hr interval that T_{int} is rising to a new regulated level, but diminish thereafter. The concept of an "adjustable thermal set-point" in thermoregulation has been challenged (3), but it is a convenient schema for interpreting the present data. Thus, the increase in O_2 consumption and working for heat, the rise in blood pressure and heart rate, and the endocrine and neuromuscular (shivering) responses are all driven by the load error between an upwardly displaced "set temperature," T_{set} and T_{int}. As T_{int} approaches T_{set}, metabolic, behavioral, endocrine, cardiovascular, and neuromuscular responses diminish proportionately towards control. Since cooling the preoptic anterior hypothalamus more deeply will not drive T_{int} higher. warmth receptors elsewhere in the body core and brain can respond to hyperthermia to limit thermogenesis

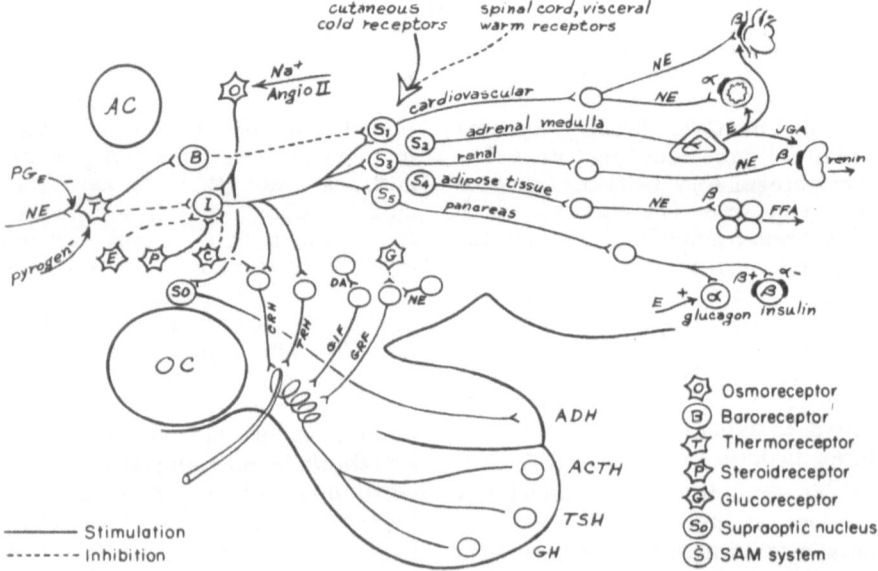

Figure 7. The schema depicts hypothetical interactions in the hypothalamus among thermal and nonthermal modalities in the regulation of body temperature (see (3) for detailed explanation).

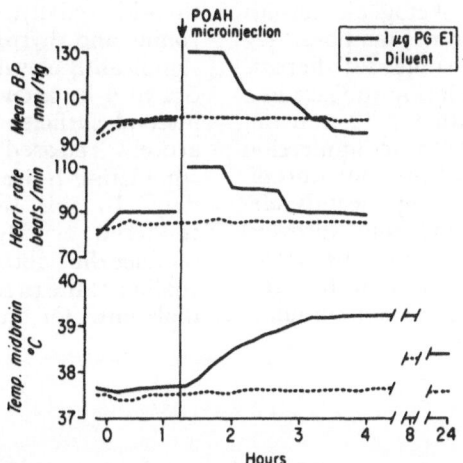

Figure 8. Elevation of blood pressure, heart rate, and internal (midbrain) temperature following microinjection of prostaglandin E_1 (1 μg) into the preoptic anterior hypothalamus (POAH) of a baboon.

and mobilize heat-loss activities. This implies that not all thermal feedback must be relayed through the preoptic anterior hypothalamus warmth detectors. The reverse of the manifold responses to preoptic anterior hypothalamus cooling is evoked by preoptic anterior hypothalamus warming: lowering of T_{int}, suppression of O_2 consumption, decline in urinary norepinephrine and epinephrine and plasma cortisol (13), inhibition of bar-pressing for warmth (7), lowering of blood pressure and heart rate (12), and fall in turnover rate of lipid substrate (8). In contrast to the constraints on thermogenic responses to preoptic anterior hypothalamus cooling, warming of the preoptic anterior hypothalamus (especially in conjunction with warming of the spinal cord) appears capable of suppressing cold defense mechanisms nearly completely. Thus, theoretically a state of torpor or even hibernation may be imposed on nonhibernating mammals, e.g., oxen (9). This is obviously a reflection of the fact that

the thermoregulatory system functions most sensitively to protect against hyperthermia, since overheating may become rapidly fatal. By comparison hypothermia, although uncomfortable, may be energetically an appropriate compromise when food and shelter are scarce, and acutely is not life threatening.

COMPARISON OF BABOON TO MAN IN ENDOCRINE RESPONSE TO COLD

The usefulness of the baboon as an experimental model for human thermoregulation is, of course, dependent on its demonstrated resemblance to the human. The similarity of the endocrine response to cold exposure in baboons and man is shown in Fig. 9. When healthy human males were immersed in 10 C water for 46 to 136 min, rectal temperature (T_r) fell a mean of 1.9 C in association with shivering, peripheral vasoconstriction and rise in metabolic rate (Hayward and Cupples, unpublished). Urinary 17-ketogenic steroids also rose with some temporal lag (perhaps due to time for hepatic metabolism) (5). Placing subjects in a 40 C warming bath for 30 min immediately after 10 C water immersion suppressed epinephrine but norepinephrine remained significantly elevated during a 145 min recovery period in 24 C T_a. The acute severe hypothermia of cold water immersion, however, did not stimulate

thyroid activity. Serum triiodothyronine and thyroxine levels were not significantly elevated in blood samples drawn 4 hr after immersion, a time interval sufficient for the effects of acutely released pituitary thyroid stimulating hormone to become evident. In this respect, the human rendered acutely hypothermic resembles the baboon subjected to local cooling of the preoptic anterior hypothalamus. Of interest, preliminary

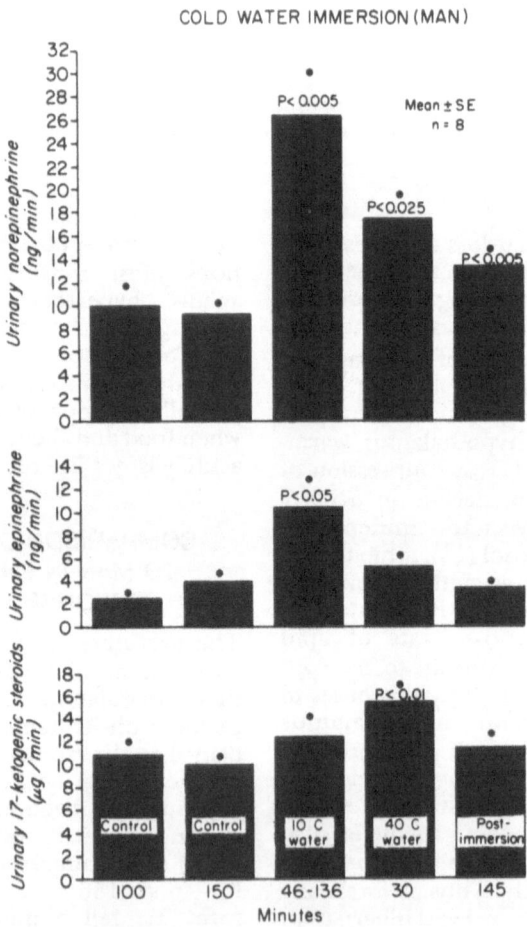

Figure 9. Rise of urinary norepinephrine, epinephrine, and 17 ketogenic steroids in human subjects immersed in 10 C water.

data suggest that urinary levels of thyrotrophin releasing hormone (the hypothalamic hormone stimulating release of thyroid stimulating hormone) rise during 10 C water immersion (5). If, indeed, thyrotrophin releasing hormone is secreted into the hypophysial portal vessels during cold water immersion, the absence of thyroidal activation may be due to concurrent release of another hypothalamic hormone which blocks pituitary response to thyrotrophin releasing hormone. Interestingly, the most recently elucidated of the hypothalamic hormones, somatostatin, which inhibits secretion of growth hormone, also has the property of blocking thyrotrophin releasing hormone-induced thyroid stimulating hormone release. It remains for future study to determine whether somatostatin is secreted in acute cold stress.

REFERENCES

1. CHAFFEE, R. R. J., AND R. J. ALLEN. *Comp. Biochem. Physiol.* 44: 1215–25, 1973.
2. EHLE, A. C., E. H. MOVGEY, F. E. WHERRY AND J. W. MASON. *Physiologist* 14: 138, 1971.
3. GALE, C. C. *Ann. Rev. Physiol.* 35: 391–430, 1973.
4. GALE, C. C., W. L. GREEN, B. R. WEBSTER AND H. SCHILLER. Jerusalem Satellite Symposia on Temperature Regulation. XXVI Intern. Congr. Physiol. Sci., 1974.
5. GALE, C. C., J. S. HAYWARD, W. L. GREEN, S. Y. WU, H. SCHILLER AND I. JACKSON. *Federation Proc.* 34: 301, 1975.
6. GALE, C. C., M. JOBIN, D. PROPPE, D. NOTTER AND H. FOX. *Am. J. Physiol.* 219: 193–201, 1970.
7. GALE, C. C., M. MATHEWS AND J. YOUNG. *Physiol. Behav.* 5: 1–6, 1970.
8. GALE, C. C., K. MURAMOTO, P. T. K. TOIVOLA AND D. STINER. *Intern. J. Biometeorol.* 15: 162–67, 1971.
9. JESSEN, C., J. A. McLEAN, D. T. CALVERT AND J. D. FINDLAY. *Am. J. Physiol.* 222: 1343–47, 1972.
10. JOY, R. T. J. *J. Appl. Physiol.* 18: 1209–12, 1970.
11. KANG, B. S., D. S. HAN, K. S. PAIK, Y. S. PARK, J. K. KIM, D. W. RENNIE AND S. K. HONG. *J. Appl. Physiol.* 26: 6–9, 1970.
12. MORISHIMA, M. S., AND C. C. GALE. *Am. J. Physiol.* 223: 387–95, 1972.
13. PROPPE, D., AND C. C. GALE. *Am. J. Physiol.* 219: 202–7, 1970.

Immunologic and morphologic effects of vasectomy in the rhesus monkey[1]

NANCY J. ALEXANDER

Oregon Regional Primate Research Center, Beaverton, Oregon 97005

Although vasectomy is the most common method of permanent male sterilization, it remains controversial because it involves surgical intervention in a healthy individual. The possibility that it may effect changes in hormone levels (6, 18), psychological changes (16, 17), and long-term systemic changes even in a small proportion of the population has been an important consideration in the controversy. Despite these fears, however, vasectomy usually does not have any adverse effects on the male psyche or hormonal milieu.

This is not to say, of course, that no changes occur after vasectomy. They do. In many animals, for example, the immune system is activated and antisperm antibodies are formed. We have been studying these immunological and morphological effects of vasectomy in rhesus monkeys as well as in man.

Antibodies to spermatozoa can be tested in various ways. We routinely use the sperm agglutination test (15) and the sperm immobilization test (12) and have followed antibody type and titer in a number of animals. For example, the test results of one group of 16 animals—9 vasectomized and 7 sham-vasectomized rhesus monkeys—have been followed for several years. Within 2 weeks after surgery, all the vasectomized animals had developed antibody levels (Fig. 1). The average spermagglutinin titer for the vasectomy group was 1:760 and two-thirds of the animals had titers above 1:640. The spermagglutinin antibody level fell almost as rapidly as it rose, and after 6 months the average antibody titer had dropped to 1:50 and no animal had a titer above 1:640. On the other hand, only one of the 7 sham-vasectomized animals developed a measurable titer of 1:40 at 2 weeks and 6 months after surgery.

Sperm-immobilizing antibodies rose and fell in a pattern similar to that of the agglutinating antibodies (Fig. 2). Two weeks after vasectomy, all the vasectomized animals had high titers; by 6 months, however, only one-third of the animals had retained these antibodies. The observation that these types of antisperm antibodies develop within 2 weeks after vasectomy indicates that antibody formation is not due to a long-term buildup of spermatozoa in the epididymis or

[1] Publication No. 789 from the Oregon Regional Primate Research Center, supported by Grants HD-4-2866, HD-05969, and RR-00163 from the National Institutes of Health.

to slow leakage from increasing back pressure. The fact that these animals developed sperm-agglutinating or sperm-immobilizing antibodies or both suggests that our tests were measuring different antibodies.

When sperm-agglutinating and sperm-immobilizing antibodies were measured in the blood of animals that had been vasectomized from 2 to 10 years, only a few had a high titer. In the 2- to 5-year postvasectomy group, only 20% had levels of sperm-agglutinating antibodies and 27% had sperm-immobilizing antibodies. Five to 10 years after vasectomy, those with sperm-agglutinating antibodies had decreased to 9%, whereas those with sperm-immobilizing antibodies remained at 27% (Fig. 3).

These agglutinating and immobilizing antibodies can be visualized indirectly by scanning electron microscopy (Fig. 4). Spermatozoa consist of a head (5 μm), midpiece, and tail, which total about 50 μm. The proximal two-thirds of the head is capped by an acrosome with an associated equatorial rim. The plasma membrane over the head is continuous and often somewhat irregular or wavy. The midpiece is surrounded by 84 to 86 mitochondria arranged in a helix of 42 to 43 gyres with 2 mitochondria for each gyre. The tail lacks mitochondria but the axonemal complex is present (24). If serum containing sperm-agglutinating antibody from a vasectomized animal is added to spermatozoa, agglutination—usually

SPERM-AGGLUTINATING ANTIBODIES

SHAM-VASECTOMIZED

Animal No.	pre-vasx	2wk.	6mo.	12mo.	20mo.	26mo.
70	-	-	-	-	-	-
27	-	-	-	-	-·	-
24	-	-	-	-	-	-
86	-	-	-	-	-	-
85	-	-	-	-	-	-
89	-	-	..	-	-	-
64	-	+	+	-	-	-

VASECTOMIZED

Animal No.	pre-vasx	2wk.	6mo.	12mo.	20mo.	26mo.
78	-	+	+	-		
80	-	++	-	-	-	-
25	-	+	-	-	-	-
21	-	++++	-	+	++	-
16	-	++++	+	-	-	-
50	-	++++	+	+	+	+
97	-	++++	-	+	++	-
91	-	++	-	-	-	-
17	-	++++	-	-	-	-

Figure 1. Results of rhesus sperm-agglutinating antibody tests after vasectomy or sham vasectomy. A titer of $0 \rightarrow 1:20 = -$; $0 \rightarrow 1:80$ $= +$; $0 \rightarrow 1:160 = ++$; $0 \rightarrow 1:320 = +++$; $0 \rightarrow 1:\geq640 = ++++$ (Emended after Alexander [3]).

SPERM-IMMOBILIZING ANTIBODIES

SHAM-VASECTOMIZED

Animal No.	pre-vasx	2wk.	6mo.	12mo.	20mo.	26mo.
70	-	-	-	-	-	-
27	-	-	-	-	-	-
24	-	-	-	-	-	-
86	-	-	-	-	-	-
85	-	-	±	-	-	-
89	-	-	-	-	-	-
64	-	-	-	-	-	-

VASECTOMIZED

Animal No.	pre-vasx	2wk.	6mo.	12mo.	20mo.	26mo.
78	-	+	-	-		
80	-	+	-	-	-	++
25	-	+	-	-	-	-
21	-	++	-	-	-	-
16	-	++	+	+	-	+
50	-	++	++	++	++	++
97	-	+	++	++	+	+
91	-	±	-	-	-	-
17	-	++	-	-	-	-

Figure 2. Results of rhesus sperm-immobilizing·antibody tests after vasectomy or sham vasectomy. $0.8-1.5 = -$; $1.5-2 = \pm$; $2-10 = +$; $10-\infty = ++$ (Emended after Alexander [3]).

of a head-to-head type—occurs (Fig. 5). Usually the flat sides of the head abut, rather than tip-to-tip. If serum containing sperm-immobilizing antibodies is added to spermatozoa, the plasma membrane is disrupted (21); this can best be seen by transmission electron microscopy (Fig. 6).

All the antibody tests mentioned so far were done with ejaculated spermatozoa. We were concerned that the antigens disclosed by our tests might have been produced by the accessory glands and might have coated the spermatozoa at the time of ejaculation. To check this possibility, we collected spermatozoa from a surgically produced draining fistula of the vas. Spermatozoa from these animals exit upon electroejaculation from the proximal vas deferens and are never exposed to accessory fluids. When these spermatozoa were diluted and used, agglutination or immobilization was similar to that of ejaculated spermatozoa, an indication that the spermagglutinins recognized in our tests were intrinsic to the spermatozoa and not coating antigens produced by the accessory glands.

We have also tested vasectomized men for antisperm antibodies and found that about 35% of them had either agglutinating or immobilizing antibodies in the serum even 10 to 25 years after vasectomy. We have not

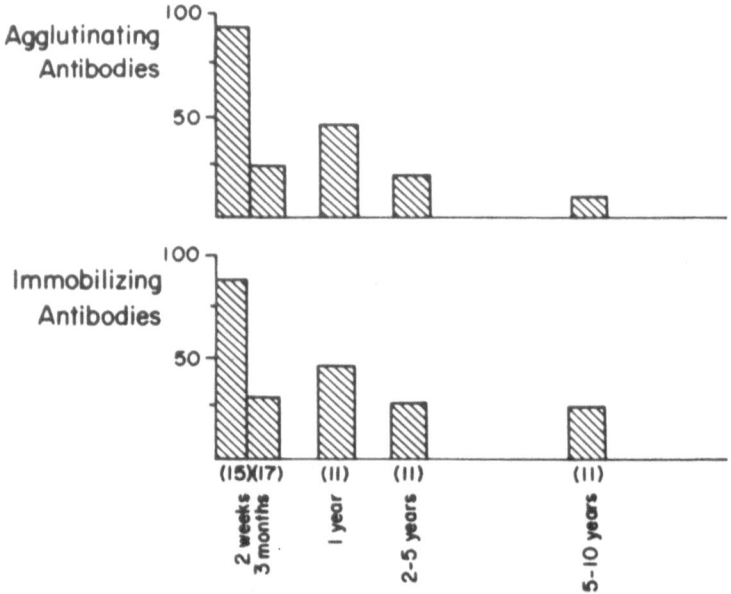

Figure 3. Percent of rhesus monkeys with antisperm antibodies after vasectomy. (n) = the number of subjects in each sample.

tested serum from many men 2 weeks to 1 month after surgery, the period during which rhesus monkeys have the highest antisperm antibody levels. We have, however, tested sera from 41 men who had had vasectomies for up to 6 months; of these, 45% had agglutinins and 32% had immobilizins. In the 5- to 10-year group, 39% had agglutinins and 38% had immobilizins. Our figure of vasectomized men with antibodies agrees with that of other workers (5, 22), although we did not find a dramatic drop in the incidence of men with sperm immobilizins 18 to 24 months after vasectomy. We did, however, find a drop in sperm immobilizins in rhesus monkeys that had been vasectomized for long periods of time: 45% in those vasectomized for 1 year or less, 27% in those vasectomized from 2 to 10 years.

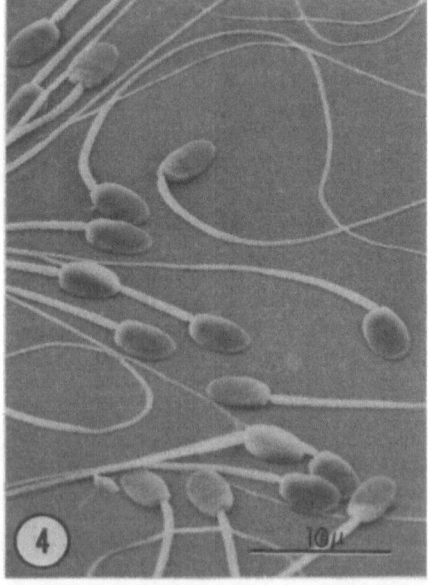

Figure 4. Scanning electron micrograph of normal rhesus spermatozoa.

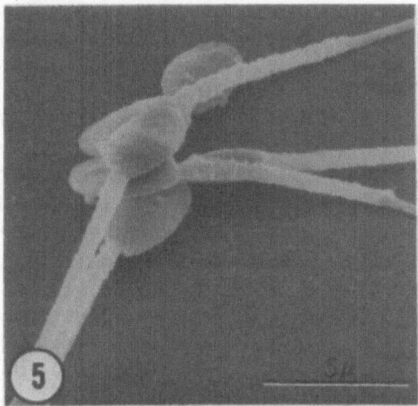

Figure 5. Scanning electron micrograph of spermatozoa incubated with serum from a vasectomized rhesus monkey and containing sperm-agglutinating antibodies. The head-to-head type of agglutination is by far the most common.

MORPHOLOGICAL CHANGES AFTER VASECTOMY

In the rhesus monkey, spermatogenesis certainly continues after vasectomy (Fig. 7), but whether daily sperm production decreases after vasectomy is not known. At any rate, millions of spermatozoa are produced daily and pass into the epididymis. Since it is important to determine how the epididymis adapts to accept these spermatozoa and how the latter are disposed of, we began a morphological study of the epithelial cells of the epididymis before and after vasectomy. We also wished to determine whether the antibodies formed after vasectomy help the body to dispose of the trapped spermatozoa.

Spermatozoa are manufactured in the seminiferous tubules of the testis. At the upper part of the rhesus testis, about 12 tubules, called efferent ducts, exit from the testis. These ducts (275 μm to 350 μm in diameter) arise from the rete and serve as conduits for the spermato-

zoa as they leave the seminiferous tubules of the testis via the rete and are propelled into the ductus epididymidis and the vas deferens.

The rete testis is lined with a simple epithelium of cuboidal to columnar cells and is thought to be the place where the fluids produced during spermatogenesis are resorbed (7). There are few surface elaborations, but occasional cells have a single cilium. The cells rest on a basal lamina (500 Å) and underlying collagen fibrils. After vasectomy, the rete remains unchanged except for a layered thickening of the basal lamina. Occasional sperm remnants can be found in the cuboidal cells (Fig. 8), but apparently vasectomy does not stimulate phagocytosis by the epithelial cells. Studies on allergic orchitis in other animals suggest that the rete is immunologically weak (13, 14), a place where leakage of sperm antigens or influxes of lymphocytes occur after vasectomy. No lo-

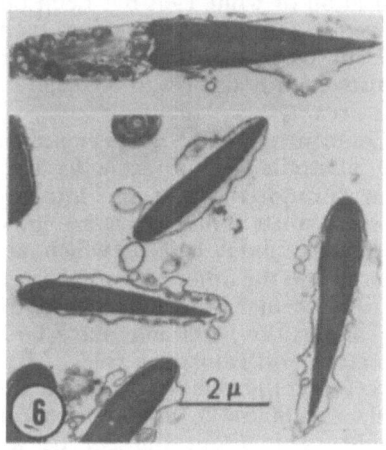

Figure 6. Transmission electron micrograph of rhesus spermatozoa incubated with serum containing sperm-immobilizing antibodies plus complement which results in lysis of the plasma membrane.

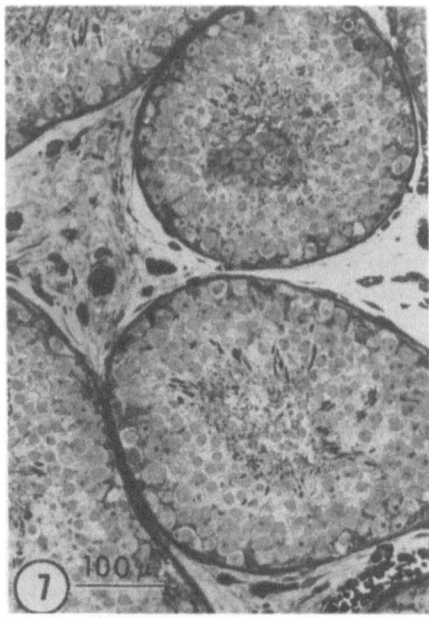

Figure 7. Light micrograph of seminiferous tubules from a previously vasectomized rhesus monkey. Normal spermatogenesis is seen.

same before and after vasectomy. Two years after vasectomy, the basal lamina has thickened, first in pleat-like folds interspersed with collagen fibrils (Fig. 9) and finally as a more solid sheet. We found no obvious morphological differences in the efferent ducts when biopsies from animals with low antibody levels were compared with those from animals with high antibody levels. All efferent ducts from vasectomized animals contained macrophages that were engulfing spermatozoa (Fig. 10). The epithelial cells themselves did not contain recognizable sperm remnants either before or after vasectomy.

The spermatozoa pass from the efferent ducts into the ductus epididymidis, a tortuous duct which is arbitrarily divided into three regions —caput, corpus, and cauda (Fig. 11). During transit, spermatozoa from the cauda become fertile and motile; unejaculated spermatozoa are stored in the cauda.

calization of white cells has been observed in samples of rete taken from the rhesus monkey; however, an exhaustive study has yet to be done on this area.

Quite incidentally, the spermatozoa are immotile and infertile as they pass through the rete and into the efferent ducts. These ducts are lined with cells, about half of which are ciliated and the other half nonciliated (1). The former propel the spermatozoa along; the latter may have both secretory and resorptive roles. After vasectomy, the cell types remain the same except for a reduction in cell height. This reduction can be ascribed to the increase in luminal content and the consequent stretching of the cells rather than to changes in cell structure. The microvilli and cilia themselves remain about the

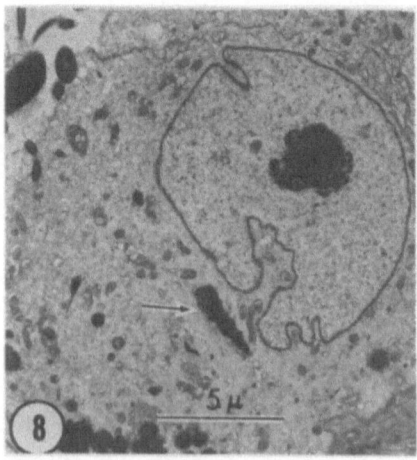

Figure 8. Transmission electron micrograph of the epithelium of the rete testis. A spermatozoan head is seen within an epithelial cell (arrow). The basal lamina is layered, an effect of vasectomy.

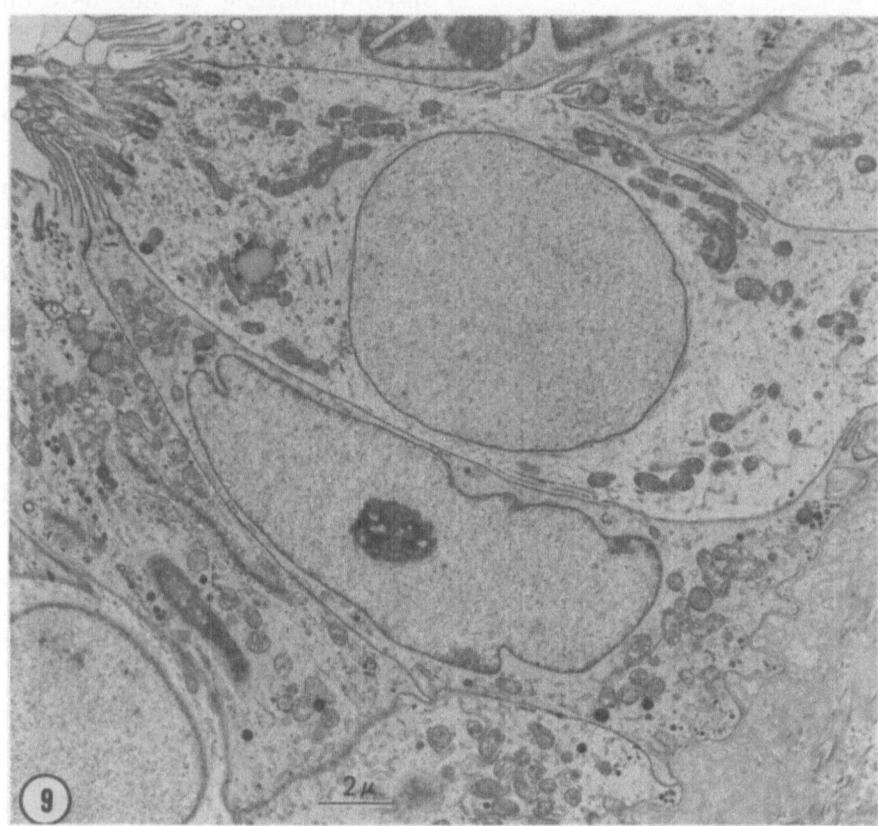

Figure 9. Part of an efferent duct from a rhesus monkey vasectomized 2 years previously. Ciliated and nonciliated cells rest upon a basal lamina which thickens after vasectomy.

After vasectomy, cells invade the epididymal epithelium and lipid-laden basal cells are seen, especially in the caput region. Throughout the epididymis, the epithelial cells retain their complement of organelles. The nuclear euchromatin and lobation, especially in the caudal regions, indicate that the cells continue to be metabolically active (11) (Fig. 12).

In fact, other than the fact that the ducts expand and become packed with spermatozoa, remarkably few morphological changes occur within the epididymal epithelium. When biopsies from animals with low antibody levels were compared with those from animals with high antibody levels, there did not seem to be much difference. Only one change appeared to correlate with antibody level. In every case except one (no. = 15), when epididymal tissue from animals with and without antibodies was compared, the presence of circulating sperm-immobilizing antibodies could be predicted by whether the spermatozoa in the lumen had broken membranes or membranes

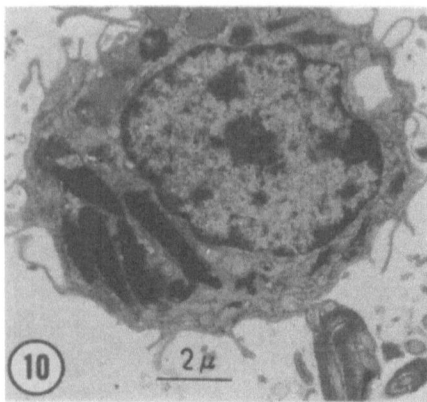

Figure 10. A macrophage containing engulfed spermatozoa from an efferent duct of an animal vasectomized 6 years previously.

that were not closely apposed to the underlying acrosome and nucleus. Apparently antibodies do get into the epididymis after vasectomy. If there is an epididymal barrier to serum proteins similar to the blood-testis barrier, it is broken after vasectomy. Studies on experimental allergic orchitis in the guinea pig support this hypothesis by showing that serum immunoglobulin can cross into the rete testis (23).

LONG-TERM EFFECTS

Another concern associated with vasectomy is whether it causes any long-term systemic dysfunctions. Scattered reports have heightened this concern, suggesting a correlation between vasectomy and some long-term detrimental effects far removed from the surgical site. For example, Roberts (19, 20) has suggested that such unexplained diseases as thrombophlebitis, arthritis, recurrent infections, multiple sclerosis, glomerulonephritis, hypoglycemia, and liver dysfunctions are associated with vasectomy. To check this possibility, we have

been conducting various tests on vasectomized animals (some of which have had vasectomies for more than 10 years) at the Oregon Regional Primate Research Center. All counts of red and white blood cells, lymphocytes, and polymorphonuclear granulocytes were within the normal range after vasectomy and did not differ significantly from those before vasectomy. When animals which had been ranked according to antibody levels as well as blood values were studied for changes after vasectomy, no significant correlations were observed.

Various blood chemistry values—such as blood urea nitrogen, creatinine, total protein, globulins, and albumin—were routinely checked to determine the health of our vasectomized animals (3, 4). At 0.5, 1, 2, 6, and 10 years after vasectomy, blood chemistry levels and, when available, mean prevasectomy values for these animals were compared with those for 40 normal males and for normal males at least 12 years old. The latter group was used to check whether changes in a particular value were associated with age. In most of these blood test results, no abnormal values or trends were detected.

Some of the blood globulin levels did change after vasectomy. Two years after vasectomy, alpha 1 globulin levels had risen significantly from the prevasectomy levels. Rank order comparisons between alpha 1 globu-

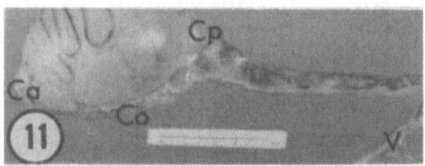

Figure 11. The testis and epididymis of a rhesus monkey. Caput (Cp), corpus (Co), and cauda (Ca) epididymis, vas deferens (V).

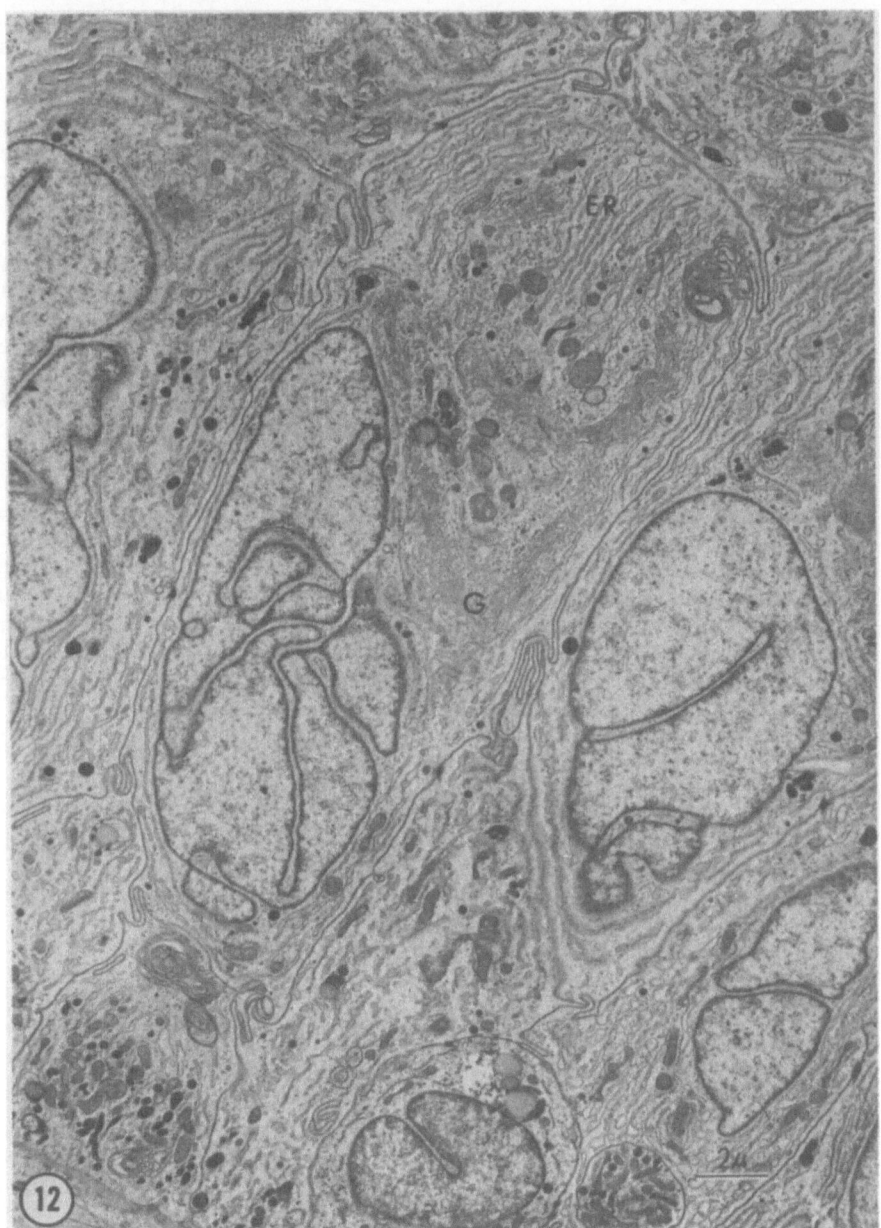

Figure 12. The epithelium of the cauda epididymis remains remarkably unchanged after vasectomy. The Golgi apparatus (G), multi-lobate nuclei, endoplasmic reticulum (ER), and numerous inclusions all attest to the activity of these cells.

lin levels and sperm antibody levels showed no correlations. Compared with those of the control group, beta globulins also rose significantly ($P < 0.001$) over the levels 2 and 10 years after vasectomy. Albumin levels dropped after vasectomy. T-tests indicate that after vasectomy, the albumin levels were significantly lower than before vasectomy ($P < 0.05$ at 6 and 10 years postvasectomy) and lower than the levels of 40 normal males ($P < 0.01$ at 6 and 10 years post-vasectomy).

The ratio of albumin to globulin (A/G ratio) in the vasectomized rhesus dropped significantly ($P < 0.001$) in the first 2 years after surgery and for several years remained lower than that in older males (Fig. 13). Normally, rhesus monkeys experience a gradual drop in the A/G ratio as they age. Because ours was a study of the same animals over a long period of time, we wished to compare their A/G ratios after vasectomy with their normal prevasectomy values and with data from normal older males (9 to 12 years of age). To determine whether some extraneous factor associated with captivity was affecting our animals, we compared the A/G ratios of the vasectomized rhesus with those of older males who had been at the Oregon Regional Primate Research Center for at least 8.5 years. This comparison yielded results similar to the previous ones, i.e., the A/G ratios of vasectomized monkeys had dropped significantly from the normal levels of older animals ($P < 0.001$).

VASOVASOSTOMY

In recent years, the sharp increase in vasectomies has been accompanied by a corresponding increase in the number of requests for vasovasostomies. The surgical success of these operations varies considerably; Dubin

and Amelar (8), for example, reported a normal sperm return that varied from 40 to 90%. High sperm counts often take months to return. Furthermore, pregnancy has been reported in only 20 to 25% of the successful cases of vasovasostomies. High spermagglutinin titers, which have been demonstrated more frequently in sterile than in fertile men (9), may cause reduced motility and agglutination of spermatozoa (10). Therefore, high antibody levels may interfere with fertility, but whether the antibody levels drop after a successful reanastomosis remains to be investigated. Moreover, whether an individual without antibodies or only one type of antibody is more likely to have fertile sperm after reanastomosis is not known.

For these reasons, antisperm antibody levels were followed in 6 rhesus monkeys after bilateral vasectomy with fulguration of the vas ends and after subsequent reanastomosis.

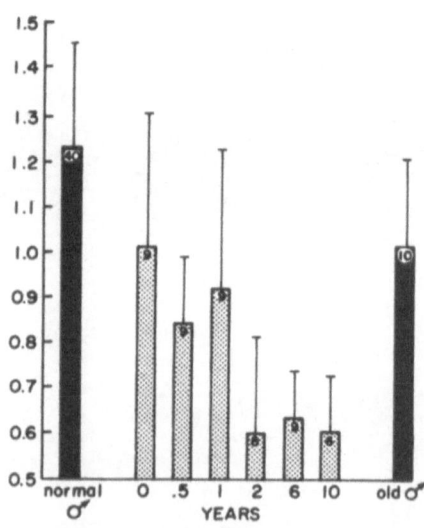

Figure 13. Comparison of the A/G ratio of normal rhesus (solid bars) and rhesus vasectomized for various lengths of time (shaded bars). no. = number within the bar.

About 10 days after vasectomy, most of the animals developed high levels of spermagglutinating antibodies, which rapidly decreased to prevasectomy levels within the next 10 days. During that time, half the vasectomized monkeys also developed complement-dependent sperm-immobilizing antibodies.

Antisperm antibody levels remained low in most of the monkeys until about 3 to 6 months after vasectomy when abrupt, seemingly spontaneous rises in sperm-immobilizing levels were observed in some of the animals but without an appreciable increase in agglutinating antibodies. It later became evident that sperm had extravasated and that the sperm granulomas that had developed could be correlated with the increase in sperm-immobilizing antibodies. Extravasated sperm in the surrounding tissue apparently induced the renewed production of immobilizing antibodies. When the animals underwent reanastomosis 6 months after vasectomy, sperm-immobilizing activity increased rapidly but in most cases subsided again in about 10 days.

The reanastomoses were surgically successful; 5 weeks postoperatively, every animal had spermatozoa in the ejaculate. Some changes were seen in the ejaculum after vasovasostomy. Rhesus semen immediately coagulates but after incubation at 37 C for 30 min, 0.6 to 1.0 ml of sperm-rich liquid is formed. For 6 months after reanastomosis, this amount averaged only 0.1 ml in our rhesus but by 8 months, it had returned to normal. In size and consistency, the rest of the plug remained unchanged when prevasectomy ejaculates were compared with those after vasectomy or after vasovasostomy.

The sperm count after vasovasostomy was about 1/3 of the prevasectomy count but improved with time. We checked the semen at about 1, 3, 6, 8, and—in some cases—12 months after vasovasostomy; on the average, the sperm count was normal by 8 months. At this time, the vasa of the animals were checked surgically and both sides were patent in every case. The fact that the count took 8 months to become normal could indicate improved passage through the vas deferens or a reduction in spermatogenesis during vasectomy. Our testicular biopsies from vasectomized animals certainly indicate that spermatogenesis continues (Fig. 7). Sperm count and the macrophages found in the ejaculum for up to 6 months after reanastomosis were not correlated with antibody level. Whether the animals had circulating antibody levels or not, agglutinated spermatozoa in the ejaculum were not seen.

Animals with high levels of sperm-immobilizing antibodies after reanastomosis also seemed to have lowered levels of sperm motility. There was no correlation between antibody levels and fertility. We are now checking these observations with a larger sample of animals. The most striking result was that when the animals with sperm-immobilizing antibody levels and sperm granulomas at the time of vasovasostomy were surgically checked again after reanastomosis, the presence of sperm granulomas at that time correlated with the presence of sperm-immobilization antibodies (2).

The author wishes to thank Jeanne Hren and Michael V. Danilchik for their excellent technical assistance.

REFERENCES

1. ALEXANDER, N. J. Vasectomy: Long-term effects in the rhesus monkey. *J. Reprod. Fertil.* 31: 399, 1972.
2. ALEXANDER, N. J. Antisperm antibodies in the rhesus after vasectomy and re-

anastomosis. Abstracts from the Seventh Ann. Mtg. Society for the Study of Reproduction, Ottawa, Canada, 1974, p. 174–175.

3. ALEXANDER, N. J. Immunologic effects of vasectomy in rhesus monkeys. PARFR Workshop on *Control of Fertility in the Male*. New York: Harper & Row, In press.

4. ALEXANDER, N. J., B. J. WILSON AND G. D. PATTERSON. Vasectomy: Immunological effects on rhesus monkeys and men. *Fertil. Steril.* 25: 149, 1974.

5. ANSBACHER, R. Vasectomy: sperm antibodies. *Fertil. Steril.* 24: 788, 1973.

6. BUNGE, R. G. Plasma testosterone levels in man before and after vasectomy. *Invest. Urol.* 10: 9, 1972.

7. CRABO, B. Studies on the composition of epididymal content in bulls and boars. *Acta Vet. Scand.* 6: 1, 1965.

8. DUBIN, L., AND R. D. AMELAR. Etiologic factors in 1294 consecutive cases of male infertility. *Fertil. Steril.* 22: 469, 1971.

9. FJÄLLBRANT, B. Sperm agglutinins in sterile and fertile men. *Acta Obstet. Gynecol. Scand.* 47: 102, 1968.

10. FJÄLLBRANT, B., and O. OBRANT. Clinical and seminal findings in men with sperm antibodies. *Acta Obstet. Gynecol. Scand.* 47: 451, 1968.

11. HAMILTON, D. W. The mammalian epididymis. In: *Reproductive Biology*. Amsterdam: Excerpta Medica Monograph, 1972, p. 268.

12. ISOJIMA, S., T. S. LI AND Y. ASHITAKA. Immunologic analysis of serum sperm-immmobilizing factor in women with unexplained sterility. *Am. J. Obstet. Gynecol.* 101: 677, 1968.

13. JOHNSON, M. H. The distribution of immunoglobulin and spermatozoal autoantigen in the genital tract of the male guinea pig: Its relationship to autoallergic orchitis. *Fertil. Steril.* 23: 383, 1972.

14. JOHNSON, M. H. An immunological barrier in the guinea-pig testis. *J. Pathol.* 101: 129, 1970.

15. KIBRICK, S., D. L. BELDING AND B. MERRILL. Methods for the detection of antibodies against mammalian spermatozoa. I. A modified macroscopic agglutination test. *Fertil. Steril.* 3: 419, 1952.

16. NASH, J. L., AND J. D. RICH. The sexual after-effects of vasectomy. *Fertil. Steril.* 23: 715, 1972.

17. PHOENIX, C. H. Sexual behavior in rhesus monkeys after vasectomy. *Science* 179: 493, 1973.

18. RESKO, J. A., AND C. H. PHOENIX. Sexual behavior and testosterone concentrations in the plasma of the rhesus monkey before and after castration. *Endocrinology* 91: 499, 1972.

19. ROBERTS, H. J. Delayed thrombophlebitis and systemic complications after vasectomy: possible role of diabetogenic hyperinsulinism. *J. Am. Geriat. Soc.* 16: 267, 1968.

20. ROBERTS, H. J. Letters to the editor. *Perspectives Biol. Med.* 14: 176, 1970.

21. RUSSO, J., AND C. B. METZ. The ultrastructural lesions induced by antibody and complement in rabbit spermatozoa. *Biol. Reprod.* 10: 293, 1974.

22. SHULMAN, S., E. ZAPPI, U. AHMED AND J. E. DAVIS. Immunologic consequences of vasectomy. *Contraception* 5: 269, 1972.

23. TUNG, K. S. K., E. R. UNANUE AND F. J. DIXON. Pathogenesis of experimental allergic orchitis. II. The role of antibody. *J. Immunol.* 106: 1463, 1971.

24. ZAMBONI, L., R. ZEMJANIS AND M. STEFANINI. The fine structure of monkey and human spermatozoa. *Anat. Rec.* 169: 129, 1971.

Index